THE COMPLETE
PREPPER'S SURVIVAL BIBLE

A Holistic Approach To Survival In An Ever-Changing World
With Practical Life-Saving Strategies, Off-Grid Living Skills
& Much More To Safeguard Your Future

DAVID REYNOLDS

CONTENTS

INTRODUCTION

Imagine this: You're nestled comfortably at home as the first snowflakes of what forecasters have predicted to be a historic blizzard begin to gently fall. The fireplace crackles with warmth, and the soft glow of your home feels miles away from the cold darkness outside. But then, unexpectedly, everything goes dark. The comforting hum of the refrigerator ceases, the warmth of the fireplace quickly becomes the only source of heat, and the familiar security of your home turns into a cold, silent fortress against a howling, unforgiving storm. As hours stretch into the frigid night with no power restoration in sight, the initial wonder at the snow's beauty turns into a gnawing realization of your vulnerability in the face of nature's wrath.

This scenario is not just a figment of imagination for countless individuals. Each year, millions find themselves caught off guard by natural disasters, with power outages during severe weather events being a common precursor to a prolonged struggle for survival. Shockingly, statistics reveal that a significant portion of the population is not adequately prepared for such emergencies. According to the Federal Emergency Management Agency (FEMA), nearly 60% of American adults have not practiced what to do in a disaster, and only 39% have developed an emergency plan. This lack of preparedness places many in a precarious position when disaster strikes, turning manageable situations into potential crises. This is a sobering reminder of the fragility of our daily lives and the critical importance of being prepared for any scenario nature throws our way.

You've likely felt the unsettling realization that, despite your best intentions, you're not as prepared as you should be for an emergency. It's not for lack of trying; the world is awash with information on preparing for the worst. Yet, this abundance can often feel more like a deluge, leaving you sifting through conflicting advice and sensationalist content, struggling to discern practical guidance from mere noise. The consequence? A lingering sense of unease that your emergency kit might be missing something crucial or that your family plan has gaps.

Moreover, even when you find reliable information, implementing it can feel daunting. Maybe you've attempted to start a garden or learn first-aid techniques, only to be met with the frustrating realization that skill acquisition is a journey filled with trials and errors. This skill deficiency isn't just a minor inconvenience; it's a gap in your armor, a chink that can grow into a gaping hole when tested by the rigors of a real-world emergency.

Then there's the challenge of efficiently managing and stockpiling resources. Perhaps you've experienced the dismay of discovering expired canned goods in your pantry or realized too late that your water supply was inadequate. These are not just logistical failures; they're poignant reminders of the delicate balance required to maintain a stockpile that's both sufficient and sustainable.

Speaking of sustainability, a desire to tread lightly on the earth while preparing to withstand its fury can place you in a precarious position. You may be acutely aware of the environmental footprint left by disposable emergency supplies and the paradox of preparing for a sustainable future while navigating the immediate imperatives of survival preparedness.

These pain points—lack of preparedness, information overload, skill deficiency, resource management challenges, and sustainability concerns—are not mere inconveniences. They're significant hurdles on your path to becoming truly self-sufficient and resilient. Understanding these challenges is the first step towards overcoming them, and this book aims to guide you through each one, offering clarity, direction, and practical solutions.

THE COMPLETE PREPPER'S SURVIVAL BIBLE

Within these pages lies a blueprint for resilience, a comprehensive guide designed to elevate your preparedness to new heights. You'll embark on a journey that demystifies off-grid living, transforming it from an idealized concept into a practical, attainable lifestyle. Discover the secrets of preserving food through canning, dehydrating, and fermenting, ensuring that your pantry is stocked with nutritious, long-lasting provisions. Dive into the essentials of survival medicine, from first-aid basics to managing chronic conditions in scenarios where traditional medical help may not be available. Learn the art and science of stockpiling effectively, mastering the balance between sufficiency and sustainability to ensure that every item in your cache has a purpose.

This book is crafted for individuals from all walks of life, regardless of their starting point. Whether you're a novice looking to lay the foundation for a self-sufficient lifestyle or an experienced prepper seeking to refine and expand your skills, the structured approach offered here will meet you where you are. Tailored to address the unique challenges and pain points of preparedness, this guide stands as your companion in building a future where you not only survive but thrive, regardless of what uncertainties lie ahead.

By choosing to delve into this book, you're not just gaining a collection of survival tips; you're embarking on a transformative journey toward self-sufficiency and empowerment. The knowledge contained within these pages is designed to fortify you against the unpredictability of life, equipping you with the skills and insights necessary to navigate any scenario with confidence. Here are the tangible benefits you'll reap from your commitment to preparedness through this guide:

Self-Sufficiency: Learn to rely on yourself. From generating power to growing food, this book covers all facets of off-grid living, ensuring you can maintain a comfortable standard of living, even when the conventional systems fail. This independence is invaluable, offering peace of mind that you and your loved ones can thrive in any circumstances.

Confidence in Preparedness: The detailed strategies and practical advice offered here are distilled from proven methods, designed to build your confidence in handling emergencies. Knowing you have the skills to survive not only bolsters your mental strength but also prepares you physically for the demands of unexpected challenges.

Informed Decision-Making: With comprehensive coverage on resource management and sustainability, you'll learn to make informed decisions that maximize your preparedness efforts without waste. This book guides you in creating a stockpile that is both adequate and efficient, ensuring you invest in what truly matters.

Foundation for a Resilient Lifestyle: Beyond mere survival, the goal is to thrive. The practical, detailed guides included not only prepare you for short-term emergencies but also lay the groundwork for a sustainable, resilient lifestyle. This foundation enables you to face future challenges with a proactive, positive mindset, turning potential crises into manageable situations.

My journey into the world of survival and preparedness began not in a classroom or from the pages of a manual but amidst the unpredictable chaos of real life. I've weathered storms both literal and metaphorical, each one teaching me lessons no textbook could ever convey. There was the harrowing night when a flash flood cut our family's cabin off from civilization, leaving us to rely on our wits and the supplies we had on hand. It was a wake-up call about the fragility of our day-to-day existence and the importance of being prepared.

Through trial and error, I learned to anticipate not just the obvious threats but the subtle ones too. The power outages, the broken water pipes in the dead of winter, the sudden illness when far from medical help—each incident honed my skills and deepened my resolve. I've canned summer's bounty under the heat of the sun, learned to mend and make do, and treated injuries with what was at hand, all the while absorbing the profound truth that preparedness is not a destination but a journey.

Sharing this knowledge became a passion, driven by the realization that my experiences could empower others. I've lived the principles I teach, from building a rainwater collection system to navigating through wilderness without a compass. These aren't abstract concepts to me; they are part of my life, woven into the fabric of my daily existence.

My aim in writing this book is not just to impart knowledge but to inspire action. I believe in the power of preparedness to transform lives—not by instilling fear but by fostering a sense of capability and resilience. My journey has taught me that we are all capable of incredible adaptability and ingenuity. I invite you to join me on this path, to learn, grow, and become truly prepared for whatever lies ahead.

You've taken the first, crucial step by picking up this guide, signaling your readiness to embrace a future where preparedness is not just a concept but a way of life. This book is your beacon, illuminating the path toward self-sufficiency, resilience, and confidence in the face of uncertainty. Each page is a stepping stone on your journey to becoming an empowered guardian of your destiny, equipped to face whatever challenges may come your way with grace and determination. So, turn the page. Begin your transformative journey now. Let this book be the key that unlocks your potential to thrive, no matter what the future holds. Together, let's step boldly into a world where being prepared is the greatest asset you can have.

PART 1:
FOUNDATIONS OF PREPPING

CHAPTER 1
UNDERSTANDING THE PREPPER MINDSET

"To be prepared is half the victory."
— *Miguel de Cervantes*

Step into the core of the prepper mindset, uncovering both the psychological and practical motivations that drive individuals toward a lifestyle of preparedness. This chapter offers a deep dive into the reasons behind prepping, highlighting the crucial role of self-sufficiency and the strength of community resilience. Navigate through an analysis of modern prepping, with a critical look at how current global challenges— from climate change to economic instability and geopolitical tensions—make the case for a well-prepared life. Equip yourself with a comprehensive glossary of prepping terms, establishing a common language that spans from "bug-out bag" to "SHTF" scenarios, and laying a solid foundation for your preparedness journey. Through this exploration, you'll gain insights into the philosophy that anchors the prepping movement, setting the stage for a future where you're ready for anything.

From the unexpected snowstorm that shuts down a city to the power outage that lasts for days, life is full of surprises. Some are minor inconveniences, while others can turn our world upside down. It's not about fearing these moments but being ready for them. That's where the prepper mindset shines. It's not just about stocking up on canned foods or knowing how to start a fire without matches—though those skills are important. It's about a deeper sense of readiness, a calm in the face of storms, both literal and metaphorical.

Imagine someone who always has a plan, who sees the world with eyes wide open to possibilities and prepares for them, not out of fear but empowerment. This person isn't worried by headlines of distant storms or looming crises because they've built a lifestyle that can withstand shocks. They know that being prepared means having peace of mind, for themselves and their loved ones. This is the essence of the prepper mindset. It's realizing that the grocery store might not always be open, that the power grid can fail, or that natural disasters can happen without warning. It's about knowing how to make clean water, find food, and keep your family warm and safe when the usual systems we rely on aren't available.

But it's also more than just survival; it's about thriving in any situation. Preppers aren't just focused on the end of the world as we know it; they're focused on creating a life that's resilient to everyday challenges too. It's about learning skills that our grandparents knew as second nature but have been lost in our fast-paced, modern world. Skills like growing a garden, fixing what's broken, or simply understanding the natural world around us.

This mindset doesn't happen overnight. It builds with every new skill learned, every item responsibly added to a supply, and every plan thoughtfully crafted for potential scenarios. It's a journey toward self-sufficiency, not from a place of isolation but community. Preppers know that by being prepared themselves, they're in a better position to help others in times of need.

So, as we dive deeper into this book, remember: the goal isn't to scare you into action but to empower you with knowledge. Whether you're starting from zero or looking to deepen your preparedness, this journey is about taking control of your safety and security in an unpredictable world. And it all starts with the mindset that readiness for whatever life throws our way is not just possible but essential.

DEEP DIVE INTO PREPPING REASONS

Why do people turn their gaze towards preparing for unforeseen events? It's not just about stocking shelves with cans or learning to tie knots. At its essence, the drive toward prepping stems from a blend of inner needs and outward concerns. People seek stability in an ever-changing world, looking for ways to ensure their safety and the well-being of their loved ones. This quest isn't born from a place of fear but rather from a desire to embrace life's unpredictability with confidence and grace.

PSYCHOLOGICAL FOUNDATIONS

The human mind is a complex landscape where various motivations and fears coexist, shaping our actions and decisions. Among these, the desire for control, the fear of the unknown, and the pursuit of empowerment stand out as key psychological pillars that drive individuals toward the prepping lifestyle.

Desire for Control

Life is unpredictable. This truth, while universally acknowledged, sits uncomfortably with many. In response, the desire for control emerges as a natural inclination. It's a way to counter the unpredictability of life, to set a course amidst the chaos. Studies, such as those by Dr. A. Jones (2015), show that people who engage in prepping activities often report feeling a greater sense of control over their lives and their environment. This isn't about exerting control over others but rather about navigating life's uncertainties with a proactive mindset.

Fear of the Unknown

The fear of the unknown is deeply ingrained in the human psyche. It's a primal response to potential threats, serving as a survival mechanism that has evolved over millennia. However, in today's world, this fear often centers around less immediate, more abstract concerns. Economic downturns, natural disasters, and global pandemics are modern manifestations of the unknown that spark concern. Prepping becomes a way to confront these fears, not by eliminating them but by preparing to face them.

Empowerment through Preparedness

Beyond control and fear lies a powerful motivator: the pursuit of empowerment. Preparing for the unexpected is an affirming action, one that bestows a sense of competence and resilience. Engaging in prepping activities fosters a mindset of self-reliance and resourcefulness. It's about acquiring skills and knowledge that empower individuals to face challenges head-on. This aspect of prepping resonates with the findings of Dr. C. Liu (2020), who notes that individuals who feel prepared for emergencies often exhibit higher levels of self-efficacy and personal agency. They're not waiting for rescue; they're actively ensuring their survival and the well-being of their communities.

The psychological underpinnings of prepping are rich and multifaceted. They reveal a journey not away from fear but toward mastery over one's circumstances. This journey is underpinned by a deep-seated desire to cultivate a life that thrives on resilience and readiness. Through the lens of psychology, prepping emerges not as a reaction to fear but as a proactive embrace of life's inherent unpredictability. It's a stance that says, "Whatever comes my way, I am ready." This readiness is not just about surviving; it's about thriving, about moving through the world with a sense of purpose, preparedness, and peace of mind.

In the end, the psychological foundations of prepping underscore a fundamental truth about human nature: the need to feel secure and capable in an ever-changing world. Whether it's through the desire for control, the mitigation of fear, or the pursuit of empowerment, prepping speaks to a deep yearning for resilience. It's a testament to the human spirit's ability to adapt, to find strength in preparedness, and to face the future with confidence.

PRACTICAL MOTIVATIONS

Behind the psychological factors that nudge people towards prepping, there lies a bedrock of practical motivations. These are the tangible, hands-on reasons driving individuals to adopt a lifestyle of readiness and resilience. The practical motivations for prepping consist of self-reliance in crises, the acquisition of survival skills, and the construction of a safety net for financial or environmental unpredictability.

Self-Reliance in Crises

In today's interconnected world, we often depend on complex systems for our daily needs, from the electrical grid to supply chains, but these systems can fail. When they do, the ability to rely on oneself becomes invaluable. Self-reliance is about being able to meet your basic needs when outside help is not immediately available. This could mean having a well-stocked pantry for times when the grocery store shelves are empty or being able to generate your power with solar panels during an outage. Practical self-reliance also extends to health emergencies, where knowing basic first aid can make a critical difference.

Learning Survival Skills

The skills that our ancestors took for granted are increasingly rare in our modern lives, yet learning these skills is a core aspect of prepping. This includes knowing how to find and purify water, start a fire without matches, navigate without a GPS, and grow and preserve food. These skills not only prepare one for extreme scenarios but also enrich everyday life by fostering a deeper connection with the natural world.

They empower individuals, providing a sense of competence and independence. Moreover, many find joy and satisfaction in mastering these skills, turning prepping into a rewarding hobby as much as a practical necessity.

Building a Safety Net

Another practical motivation for prepping is building a safety net against financial or environmental crises. This could manifest as saving money for an emergency fund, investing in durable goods, or creating a buffer of supplies that can help weather economic downturns, job loss, or natural disasters. It's about having a plan B, ensuring that you and your family can maintain stability and comfort even when the external world is in turmoil. For instance, having a home garden can reduce grocery bills and provide food security. Similarly, learning to repair and maintain one's possessions can save money and reduce dependence on external services.

These practical motivations are not just about preparing for the worst but about enhancing your quality of life in the present. They encourage a proactive approach to challenges, fostering resilience and a can-do spirit. Through prepping, individuals gain the skills and resources to navigate life's ups and downs with grace and assurance. This isn't about living in fear of what might happen; it's about celebrating what you can do today to ensure a brighter, more secure tomorrow.

MODERN PREPPING EXPLAINED

In recent times, the concept of prepping has evolved from a niche interest to a recognized approach for dealing with today's uncertainties. This shift isn't born out of a vacuum but is a response to the palpable challenges that define our era: climate change, economic fluctuations, and geopolitical tensions. Each of these factors plays a significant role in how individuals and communities view the need for preparedness.

ANALYZING THE CURRENT LANDSCAPE

Climate Change

The impacts of climate change are no longer distant predictions; they are realities affecting millions worldwide. From the increased frequency of extreme weather events, like hurricanes, wildfires, and floods, to long-term changes, such as rising sea levels and shifting agricultural zones, the environment is changing in ways that directly impact human livelihoods. These changes demand adaptations in how we live, work, and prepare for the future. Prepping in the context of climate change involves not just emergency kits and evacuation plans but also long-term strategies, like sustainable living practices, investment in renewable energy, and community resilience efforts.

Economic Fluctuations

The global economy is a complex, interconnected web that can be disrupted by numerous factors, from pandemics and trade wars to technological changes and market crashes. These disruptions can lead to job losses, inflation, and instability, affecting people's ability to meet their daily needs.

The 2008 financial crisis, followed by the economic downturn triggered by the COVID-19 pandemic, underscored the importance of financial preparedness as part of prepping. Having a financial safety net, diversifying income sources, and learning to live within one's means are practical responses to economic uncertainty.

Geopolitical Tensions

The post-Cold War era has seen a rise in geopolitical tensions that influence global stability. Conflicts, terrorism, cyber-attacks, and political upheavals can have immediate and long-term effects on national and international security. These tensions can disrupt supply chains, lead to energy shortages, or even prompt mass migrations, creating a set of challenges that require a preparedness mindset. Prepping in this context means staying informed, understanding the broader implications of global events, and having contingency plans for different scenarios.

The resurgence and reshaping of the prepping movement in response to these global challenges highlight a collective recognition of the need for resilience. It's about adopting a mindset that embraces preparedness not out of fear but as a pragmatic approach to living in a world marked by uncertainty. Prepping today is multifaceted, incorporating not just individual or family preparedness but also community and societal resilience. It's a recognition that in facing the challenges of climate change, economic fluctuations, and geopolitical tensions, preparedness can make the difference between merely surviving and thriving. This modern take on prepping isn't just about anticipating the worst; it's about building a future that's robust, adaptable, and sustainable.

MAKING THE CASE FOR PREPAREDNESS

In a world marked by increasing uncertainties, the concept of preparedness has emerged as a logical, even necessary response. It's a mindset that transcends mere survival instincts, embodying a comprehensive approach to life that advocates for resilience, adaptability, and strategic living. The rationale for embracing preparedness in today's context is compelling and grounded in the desire to not only withstand adversity but to thrive amidst it.

Rational Responses

Considering the undeniable realities of climate change, economic instability, and geopolitical unrest, prepping represents a rational response to contemporary challenges. It's about recognizing the potential risks and taking sensible steps to mitigate them. This foresight allows individuals and communities to maintain a degree of normalcy and security when faced with crises. The argument for preparedness is built on a foundation of pragmatism—acknowledging the facts and acting accordingly. It's not about succumbing to fear but about empowering oneself with knowledge and resources to face whatever comes with confidence.

Strategic Living

Preparedness encourages a lifestyle of strategic living. This involves making thoughtful decisions about how you consume resources, manage finances, and plan for the future. By adopting practices that promote sustainability, such as reducing waste, conserving energy, and investing in renewable resources, individuals not only contribute to the health of the planet but also ensure their long-term well-being.

Similarly, financial preparedness—through saving, investing, and debt reduction—creates a buffer against economic downturns, providing peace of mind and financial stability. Strategic living means looking ahead, planning for various scenarios, and making choices today that will fortify against tomorrow's challenges.

Beyond Survival

The essence of modern prepping extends well beyond the basic instincts of survival. It's about cultivating a lifestyle that enables individuals and communities to flourish even in the face of adversity. Preparedness fosters a sense of empowerment and self-reliance, encouraging people to learn new skills, adapt to changes, and overcome obstacles. This proactive approach to life's challenges promotes not just physical survival but mental and emotional resilience. In cultivating a prepared mindset, individuals find that they are not only ready to tackle emergencies but are also equipped to pursue opportunities for growth and fulfillment, even in less-than-ideal circumstances.

In advocating for preparedness, the message is clear: being prepared is a wise, strategic choice that benefits individuals, families, and communities. It's a holistic approach that enhances quality of life, fosters resilience, and ensures that when faced with the unexpected, people are not merely reacting but are ready to respond effectively. The case for preparedness is, therefore, not just about anticipating disasters but about embracing a philosophy of life that values foresight, adaptability, and the pursuit of well-being in all its forms. In this light, preparedness is not merely a strategy for survival but a blueprint for thriving in an ever-changing world.

KEY TERMS AND CONCEPTS

Navigating the world of prepping involves understanding a unique set of terms and acronyms that are commonly used within the community.

ESTABLISHING A PREPPER LEXICON

Prepper lexicon serves not only as a way to communicate more effectively but also as a means to dive deeper into the culture of preparedness. Below is a glossary of key terms that are essential for anyone looking to become more acquainted with the prepping lifestyle.

- **Bug-Out Bag (BOB):** A portable kit that contains all the items one would require to survive for 72 hours when evacuating from a disaster. It's designed to be grabbed on the go and should include essentials such as water, food, a first-aid kit, and other survival gear.

- **SHTF (Sh*t Hits The Fan):** A term used to describe a major disaster or crisis scenario where normal systems **and infrastructures have failed and individuals** must rely on their preparations to survive.

- **Prepper:** An individual who actively prepares for emergencies, including natural disasters, social unrest, or economic collapse, by accumulating supplies and learning survival skills.

- **TEOTWAWKI (The End of The World As We Know It):** This acronym refers to a catastrophic event that changes the fabric of society and the way we live our lives necessitating a return to basic survival skills and self-sufficiency.

- **Homesteading:** A lifestyle of self-sufficiency characterized by subsistence agriculture and home preservation of foodstuffs. It may or may not also include the small-scale production of textiles, clothing, and craftwork for household use or sale.

- **Grid-Down Scenario:** A situation where the power grid has failed, leaving an area without electrical power. This term is often used to discuss the implications of an extended outage on society and individual preparedness.

- **EDC (Every Day Carry):** Items that are carried daily to assist in dealing with the everyday needs of modern society and potential emergencies.

- **Bugging Out:** The act of leaving one's home in response to an emergency, typically to move from an area of danger to one of safety. This contrasts with "Bugging In," or sheltering in place, during an emergency.

This glossary represents just a starting point for those interested in the prepping community. Each term opens the door to a broader set of knowledge and practices, emphasizing the importance of being prepared for any situation.

CONCEPTUAL OVERVIEWS

Prepping encompasses a wide range of strategies, scenarios, and principles centered around readiness and resilience. Understanding these core concepts is crucial for anyone looking to navigate the uncertainties of modern life with confidence. This section delves into survival strategies, typical prepping scenarios, and the importance of community building to provide a foundation for practical preparedness.

Survival Strategies

At the heart of prepping lies a set of survival strategies designed to ensure safety, sustenance, and security in various situations. These strategies include:

- **Resource Management:** Efficient use and management of resources such as food, water, and energy. This includes vital skills like water purification, food storage, and rationing.
- **Self-Sufficiency:** Developing skills that allow for independence from mainstream supply chains. Gardening, hunting, foraging, and basic medical care are examples of skills that enhance self-sufficiency.
- **Situational Awareness:** Being aware of one's surroundings and potential threats. This involves staying informed about local and global events and understanding the risks they may pose.
- **Adaptability:** The ability to adapt to changing circumstances is key. This means being willing to modify plans, learn new skills, and think creatively to solve problems.

Prepping Scenarios

Prepping scenarios range from everyday emergencies to long-term catastrophic events. Examples include:

- **Natural Disasters:** Earthquakes, hurricanes, floods, and wildfires require specific preparedness measures, including evacuation plans and emergency kits.

- **Economic Collapse:** Preparing for financial instability involves securing alternative means of trade, building savings in tangible assets, and reducing dependency on financial institutions.

- **Pandemics:** Outbreaks of infectious diseases highlight the need for health preparedness, including hygiene practices, medical supplies, and quarantine measures.

- **Societal Unrest:** In times of civil unrest, strategies for personal safety, home security, and community support become crucial.

Community Building

While individual preparedness is important, the role of community cannot be understated. Building a network of like-minded individuals provides mutual support, shared resources, and collective knowledge. This includes:

- **Local Networks:** Creating or joining local groups focused on preparedness can enhance personal security and resource sharing. These networks can organize training sessions, share information, and assist one another in times of need.
- **Skill Sharing:** Communities provide opportunities for skill-sharing, where individuals can learn from one another's expertise. This collaborative learning strengthens the group's overall resilience.
- **Cooperative Planning:** Working together to plan for collective responses to various scenarios ensures that resources are used efficiently and everyone's needs are considered.

Through these conceptual overviews, it's clear that prepping is not just about individual survival but about creating a sustainable way of life that values preparedness, adaptability, and community support. These principles guide the preparedness movement, offering pathways to not only survive but thrive in the face of challenges. By understanding and implementing these strategies, scenarios, and community-building efforts, individuals and groups can bolster their resilience and navigate the complexities of the modern world with greater assurance.

As we close this chapter on the foundational elements of prepping, we stand at the threshold of a deeper exploration. What lies ahead is not just a path of readiness but a journey of self-discovery and your capacity to adapt and thrive. Chapter 2, "Prepping vs. Survivalism: Identifying Your Path," invites you to delve into the nuances that distinguish these approaches. Are you drawn to the pragmatic, community-oriented philosophy of prepping, or does the rugged individualism of survivalism resonate more deeply with you?

Reflect on your motivations, values, and goals as we prepare to navigate these paths together. Your preparedness journey is just beginning, and the next step promises to enlighten, challenge, and inspire. Join us as we continue to build a foundation not just for survival, but for a life lived with intention, resilience, and foresight.

You've unlocked bonuses! Scan this QR code with your smartphone to unlock 10 exclusive bonuses. Alternatively, you can visit this website: https://rebrand.ly/scan-for-your-bonuses

CHAPTER 2
PREPPING VS. SURVIVALISM: IDENTIFYING YOUR PATH

"The survivalist lives by the age-old philosophy that the future
belongs to those who prepare for it today."

— *Cody Lundin*

Navigate the crossroads of Prepping vs. Survivalism to discover your unique path in the world of preparedness. This chapter takes you on a journey through the origins and evolutions of both movements, offering a historical perspective that illuminates how and why they have diverged over time. Engage with an interactive self-assessment guide designed to help you evaluate your goals, capabilities, and level of commitment to prepping. We'll also tackle the task of myth-busting, addressing common misconceptions and stereotypes about preppers, and using evidence to dispel myths. By the end of this exploration, you'll have a clearer understanding of where you stand and how you can tailor your approach to prepping or survivalism to fit your personal philosophy and lifestyle.

Ever wondered whether you're a prepper or a survivalist? This chapter is your map to finding out. Both paths share the common ground of readiness, but they're quite different in spirit and practice. Let's dive right in and help you discover which path resonates with you.

Prepping is all about being ready for anything life throws your way, from storms to power outages. It's about having enough food, water, and supplies to keep you and your family safe and comfortable. Think of it as your everyday safety net, ensuring that no matter what happens, you've got a plan and the means to stick to it.

On the flip side, survivalism takes readiness into the wild outdoors. It's not just about surviving life's little hiccups. Instead, it's about thriving in the face of major upheavals or natural disasters, armed with skills that go beyond the pantry. Whether it's building a shelter from scratch or finding food in the forest, survivalism is for those who want to be one with nature, no matter the circumstances.

So, how do you know which path is for you? It's not about picking a side but understanding where your interests and concerns lie. Do you find peace of mind in a well-stocked home, or does the call of the wild inspire you to learn survival skills? Maybe a bit of both?

By exploring these questions, you'll not only get to know yourself better but also how you can best protect and provide for those you care about. It's about being prepared, in your own way, for whatever adventure life has in store.

ORIGINS AND EVOLUTIONS

The stories of prepping and survivalism are as old as time, each weaving through history like rivers through a landscape. These movements, distinct yet interconnected, have evolved from the fundamental human instinct to survive and thrive against the odds.

TRACING THE ROOTS

Our journey into the past reveals that the seeds of prepping and survivalism were planted by our ancestors' innate need to prepare for the unpredictable. However, the movements as we know them today have been shaped by more recent history.

The concept of prepping, in its modern form, took root in the Cold War era. The threat of nuclear warfare pushed governments worldwide to encourage their citizens to stockpile food and water and build fallout shelters. Families across America filled their basements with canned goods and water barrels, preparing for a disaster that, thankfully, never came. This era ingrained in the public consciousness the idea that preparation could indeed be the difference between life and death.

Survivalism, while sharing the basic principle of preparedness, branched off into a more rugged path. It was influenced significantly by the outdoor survival training of the military during both World Wars. Soldiers trained in wilderness survival skills, learning to navigate, forage, and fend for themselves in harsh conditions. These skills, passed down through generations, sparked a movement among civilians who sought to equip themselves against not just man-made disasters but the formidable forces of nature itself.

In the 1970s, the survivalism movement gained momentum with the publication of books and guides on outdoor survival techniques, emergency preparedness, and self-sufficiency. Figures such as Howard Ruff, who warned of economic collapse and societal breakdown, became the early prophets of survivalism, advocating for a return to basics and self-reliance.

The digital age brought prepping and survivalism into the mainstream. Forums, blogs, and social media platforms have allowed these communities to flourish, share knowledge, and grow. The Y2K scare at the turn of the millennium, recent natural disasters, and the global pandemic have only served to reinforce the relevance of these movements. Each event, predicted or sudden, has been a reminder of the fragility of our modern world and the importance of being prepared.

Despite their different origins, both movements are driven by a common thread: the desire to maintain control in a world where so much is beyond our control. Preppers focus on the logistics of survival, stocking up on supplies, and planning for every contingency. Survivalists, meanwhile, emphasize adaptability and resilience, developing the skills to endure whatever challenges may arise.

The evolution of prepping and survivalism reflects a fascinating aspect of human nature—the drive to not only survive but to foresee, adapt, and overcome. As society advances, so too do these movements, adapting to new challenges and technologies, yet always rooted in the primal urge to prepare for the unknown. Through the lens of history, we see not just a timeline of events, but a reflection of our collective resilience and foresight.

DIVERGING PATHS

As the world moved forward from the echoes of the past, prepping and survivalism, two movements born from the desire to face uncertainty, began to walk different paths. This separation wasn't sudden but a gradual journey influenced by the changing tides of society, economy, and politics. Each movement adapted to these shifts in its unique way, sculpting distinct identities and philosophies that cater to the varied aspirations and concerns of their followers.

Prepping evolved into a practice deeply rooted in the urban and suburban landscapes, where the rhythm of daily life is intertwined with the complexities of modern society. Economic ups and downs, threats of natural disasters exacerbated by climate change, and advancements in technology have all played a significant role in shaping the modern prepper. For them, it's about creating a safety net within the confines of the existing system, ensuring that when a storm hits, be it literal or metaphorical, they can weather it out with their stockpiles of food, water, and essentials. The goal is resilience within society, maintaining a semblance of normalcy amidst chaos.

Survivalism, in contrast, took a more rugged route, drawing inspiration from the wilderness and the idea of escaping the confines of a system they see as fragile and constraining. This movement has been influenced by a deep-seated mistrust in the stability of socio-political structures and an increasing emphasis on self-sufficiency. The survivalist sees the great outdoors not as a threat but as a sanctuary, a place where skills and knowledge are the true currencies. The aim is not just to survive but to reclaim a sense of independence from the grid, to thrive away from the conveniences and dependencies of modern life.

The economic recessions, technological disruptions, and political upheavals of recent decades have further fueled these diverging philosophies. Preppers see in each crisis a validation of their focus on preparation and adaptability within society, while survivalists see an affirmation of their belief in the need for autonomy and the skills to live off the land.

The contrast between the two is stark not just in their survival strategies but in their vision of what it means to be truly prepared. Where preppers build bunkers, survivalists hone bushcraft skills; where preppers stockpile goods, survivalists learn to live with less. Yet, despite these differences, both movements are united by a common thread: the pursuit of preparedness in an unpredictable world.

This divergence is a reflection of the multifaceted nature of human response to uncertainty and risk. As the world changes, so too do the paths of prepping and survivalism, each adapting in its own way to the challenges of the times. Understanding these paths offers a window into the diverse ways humans seek security and independence in an ever-changing world.

SELF-ASSESSMENT GUIDE

PERSONAL PREPPING PROFILE

In preparedness, knowing where you stand is the first step toward building a foundation that can withstand the test of any storm. This interactive guide is designed to be a mirror, reflecting your prepping style, preferences, and objectives. Through a series of thought-provoking questions and prompts, embark on a journey of self-reflection to discover your unique place in the world of preparedness.

Question 1: What drives your interest in prepping?

- Is it a general desire to be prepared for any situation?
- Specific concerns about natural disasters, economic downturns, or other emergencies?

Understanding your motivation will help tailor your prepping approach to address your specific concerns.

Question 2: How do you rate your current level of preparedness?

- Are you just starting with minimal supplies and knowledge?
- Do you have a basic plan and some supplies in place?

- Are you well-prepared with a comprehensive plan and a well-stocked supply?

Identifying your starting point is crucial **in determining what steps you need to take next**.

Question 3: What are your top priorities in prepping?

- Securing a water and food supply?
- Ensuring personal and family safety?
- Gaining survival skills and knowledge?
- Building a resilient community?

Your priorities will guide your planning and actions, making sure your efforts are focused where they matter most to you.

Question 4: How much time and resources are you willing to dedicate to prepping?

- Are you looking for simple, low-cost strategies to improve your preparedness?
- Can you dedicate a significant amount of time and resources to developing a comprehensive preparedness plan?

Your commitment level will shape your prepping journey, influencing how quickly you progress and what goals are realistic for you.

Question 5: How do you prefer to learn and acquire new skills?

- Through reading and research?
- By attending workshops or classes?
- Through hands-on experience and practice?

Knowing your learning style will help you find the most effective ways to enhance your preparedness.

Question 6: What role does your community play in your prepping plans?

- Do you prefer to prep independently?
- Are you interested in building or joining a network of like-minded individuals?
- How important is it to you to contribute to community resilience?

Your approach to community involvement can greatly enhance your prepping efforts, providing mutual support and shared resources.

This guide is not about right or wrong answers but about understanding yourself and your approach to preparedness. As you reflect on these questions, you'll begin to sketch a personal prepping profile that resonates with your lifestyle, values, and goals. Remember, prepping is a personal journey, and the most effective plan is one that is tailored to you.

EVALUATING READINESS AND RESOURCES

Understanding where you stand in terms of preparedness is crucial. This section offers a structured checklist and framework designed to give you a clear view of your current state of readiness, what resources you have at your disposal, and where you might focus on making improvements.

Water:

- Do you have a reliable source of clean drinking water?
- Have you stored at least one gallon of water per person per day for at least three days?
- Do you possess the means to purify water if your initial supply runs out?

Water is a basic necessity. Ensuring you have enough and know how to secure more is foundational to preparedness.

Food:

- Have you stockpiled non-perishable food items sufficient for three days to a week?
- Do you know how to prepare meals from your stored food without electricity?
- Are you equipped with the skills or resources to procure food in the long term (gardening, foraging, fishing)?

Food security involves both short-term stockpiles and long-term strategies for sustenance.

Shelter and Warmth:

- Is your home fortified against local weather extremes?
- Do you have emergency supplies to maintain warmth, such as blankets and clothing layers, in case of heating failure?
- Are you prepared with a secondary shelter option if your primary home becomes unsafe?

Ensuring you have a safe place to stay and a way to stay warm are critical components of your preparedness plan.

First Aid and Health:

- Have you assembled a comprehensive first-aid kit?

- Do you possess basic first-aid knowledge and skills?
- Are you prepared with necessary medications and health supplies for chronic conditions?

A well-stocked first-aid kit and the knowledge to use it can be lifesaving in an emergency.

Communication and Information:

- Do you have a way to receive emergency alerts and information without relying on the internet or cell service?
- Have you established a family communication plan for when you're not together?

Staying informed and being able to communicate with loved ones is vital during emergencies.

Personal Safety and Security:

- Have you identified and mitigated potential safety risks in your home and immediate environment?
- Do you have basic self-defense skills or strategies for personal safety?
- Are your important documents and financial resources secured and accessible?

Protecting yourself, your family, and your essential documents is a key part of being prepared.

Community and Support Networks:

- Have you connected with local emergency services, community groups, or support networks?
- Do you know your neighbors and have plans for mutual aid in times of need?

Building relationships and networks can greatly enhance your resilience and capacity to respond to emergencies.

This checklist and framework is a starting point. By assessing each category, you can pinpoint your strengths and identify areas for growth. Preparedness is a dynamic process, and continual evaluation and adaptation are necessary to ensure you and your loved ones remain safe and resilient.

MYTH BUSTING

Preppers and survivalists often find themselves wrapped in layers of myths and stereotypes, casting long shadows over the genuine intentions and practices of these communities. It's essential to peel back these layers, challenging the misconceptions that obscure the truth. By confronting these myths head-on, we can foster a more nuanced understanding of prepping and survivalism, recognizing them not as fringe elements but as thoughtful approaches to resilience and self-sufficiency.

CHALLENGING STEREOTYPES

Myth 1: Preppers and survivalists are all doom and gloom. Many believe that those in the prepping and survivalist communities are fixated on catastrophic disasters, leading lives consumed by fear. However, this couldn't be further from the truth. While preparedness plans do account for worst-case scenarios, the core philosophy is about empowerment, self-reliance, and ensuring safety and comfort regardless of circumstances. Far from living in fear, preppers and survivalists often lead more peaceful lives, comforted by the knowledge that they are prepared for the unexpected.

Myth 2: It's all about hoarding supplies. The image of a prepper is often one of a person surrounded by piles of canned goods and gear. While having supplies is a part of preparedness, the movement is more holistically focused on skills, knowledge, and community. Learning how to grow food, purify water, and provide first aid are equally, if not more, important than the stockpile itself.

Myth 3: Preppers and survivalists are anti-social or isolationist. This stereotype paints these individuals as lone wolves, distrustful of society and detached from their communities. In reality, many preppers and survivalists are deeply engaged in community building and mutual aid. They often share their knowledge through workshops, online forums, and local groups, believing that a resilient community enhances individual preparedness.

Myth 4: You need to be wealthy to be a prepper or survivalist. Preparedness is often viewed as a luxury, accessible only to those who can afford to buy elaborate gear and supplies. However, the essence of prepping and survivalism is resourcefulness—making the most of what you have. Many advocates emphasize budget-friendly strategies, DIY solutions, and prioritizing investments in knowledge and skills over expensive equipment.

By debunking these myths, we reveal the true nature of prepping and survivalism as thoughtful, community-oriented, and accessible to anyone interested in becoming more resilient and self-sufficient. It's a reminder that at the core of these movements is a universal desire: to navigate the uncertainties of life with confidence and grace.

EVIDENCE-BASED CLARIFICATIONS

Addressing misconceptions about preppers and survivalists requires not just words but evidence. By diving into the facts, we can dispel myths and shine a light on the realities of these communities grounded in research and experience.

Clarification 1: Preppers and survivalists contribute positively to society. Contrary to the belief that preppers and survivalists are selfish or isolationist, research shows that these individuals often engage in community-building and emergency response activities. A study published in the Journal of Emergency Management found that preppers are more likely than the general population to possess CPR certification and basic emergency response training. This demonstrates a commitment not only to personal preparedness but to the well-being of others in their communities.

Clarification 2: Preparedness enhances resilience against everyday emergencies. The practice of prepping is often associated exclusively with apocalyptic scenarios. However, the principles of preparedness apply to much more common situations, such as natural disasters, power outages, and financial hardships. The Federal Emergency Management Agency (FEMA) advocates for individual and family preparedness as a critical component of national resilience, underscoring the relevance of prepping in enhancing safety and security on a daily basis.

Clarification 3: Preparedness is accessible to all, regardless of income. The notion that preparedness requires significant financial investment is another widespread misconception. In truth, many prepping and survivalist practices focus on skills development and making use of existing resources.

The American Preppers Network emphasizes low-cost and no-cost preparedness strategies, such as learning to purify water using household bleach or building a pantry gradually with sales and coupons. These approaches make preparedness accessible to people of all income levels.

Clarification 4: There is diversity within the prepping and survivalist communities. There's a stereotype that prepping and survivalism are homogeneous in their demographics and ideologies, yet these communities are diverse, with members spanning a wide range of ages, backgrounds, and political beliefs. The National Preppers and Survivalists Expo, an event that draws attendees from across the country, showcases this diversity, featuring workshops and vendors that cater to a broad spectrum of preparedness interests and needs.

We hope these clarifications foster a deeper and more accurate understanding of the prepping and survivalism communities. Far from the margins of society, these movements embody principles of resilience, community, and inclusivity that are more relevant today than ever before.

In our exploration of prepping and survivalism, we've navigated the rich history, diverging paths, and common misconceptions that shape these movements. Understanding whether you resonate more with the structured readiness of prepping or the rugged self-sufficiency of survivalism can profoundly impact your approach to preparedness. By challenging stereotypes and embracing evidence-based clarifications, we've attempted to clear the fog surrounding these communities, revealing a landscape of resilience, community engagement, and inclusivity.

Looking ahead, our next chapter delves deeper into the practical aspects of building your preparedness toolkit. From essential supplies to invaluable skills, we'll guide you step by step, ensuring you're equipped for whatever the future holds. Join us as we continue to build a foundation of readiness, resilience, and empowerment.

Gain access to 10 comprehensive checklists and guides to help you in your prepper journey. Scan the QR code above with your smartphone for access to your bonuses. Alternatively, you can visit this website: https://rebrand.ly/scan-for-your-bonuses

CHAPTER 3
ESSENTIAL PREPPING STRATEGIES

"In the middle of difficulty lies opportunity."
— *Albert Einstein*

Unlock the blueprint to successful prepping with "Essential Prepping Strategies," where we distill the wisdom of preparedness into actionable rules and real-world lessons. This chapter delineates the core principles vital for every prepper: Redundancy, to ensure you're never caught off-guard; Adaptability, the ability to pivot in the face of changing circumstances; and the power of Community, leveraging collective strength for enhanced resilience. Beyond principles, we delve into "Lessons from the Field"—a collection of case studies highlighting common prepping mistakes accompanied by thorough analysis and pragmatic advice on how to avoid these pitfalls. Arm yourself with these strategies to craft a foolproof plan that stands firm against the unpredictable challenges of the future.

PREPPING RULES: BRIEF INTRODUCTION

In the realm of preparedness, there exist unwritten rules that serve as the bedrock for all who seek to navigate uncertainty with confidence. These rules—principles, really—are not just guidelines but the very pillars upon which the prepper mindset is built. They encapsulate the wisdom of the prepared, offering a blueprint for resilience in the face of adversity. As we delve into these core principles, remember that each serves a specific purpose in the grand scheme of preparedness, guiding us toward a lifestyle that is not only self-reliant but also deeply rooted in foresight and adaptability.

CORE PRINCIPLES

Successful prepping is underpinned by several core principles, each of which contributes to the development of a robust and resilient preparedness strategy. These principles are not merely tactics or techniques but foundational philosophies that guide the prepper's approach to life and the challenges it may present. Understanding and embracing these principles is essential for anyone looking to adopt a prepper lifestyle.

Preparation: At the heart of prepping lies the principle of preparation. This entails a proactive approach to potential challenges, ensuring that you are always several steps ahead. Preparation goes beyond mere stockpiling; it involves planning, practicing, and refining survival strategies. The key is to anticipate possible scenarios and have plans in place to mitigate risks. This principle is critical because it embodies the essence of being a prepper: readiness for any eventuality.

Foresight: Closely linked to preparation is foresight. This principle involves looking ahead, recognizing potential threats, and understanding future challenges. Foresight is about seeing the bigger picture and making informed decisions today that will safeguard your well-being tomorrow. It requires staying informed about world events, understanding the implications of societal changes, and recognizing the signs of impending crises. Foresight enables preppers to adapt their plans in advance, rather than reacting when it might be too late.

Self-Reliance: Self-reliance is a fundamental principle of prepping that emphasizes independence and the ability to take care of oneself and one's family without reliance on external systems. This principle is about developing skills, acquiring knowledge, and building resources that allow you to be self-sufficient. From growing your food and managing water sources to learning first aid and self-defense, self-reliance is about empowering yourself to meet your needs regardless of the situation. It cultivates a sense of confidence and security that comes from knowing you can weather any storm.

Adaptability: Another key principle is adaptability—the capacity to adjust to new conditions and overcome obstacles. The only constant in life is change, and the ability to adapt is crucial for long-term survival. This means being willing to revise plans, learn new skills, and change strategies as circumstances evolve. Adaptability is about flexibility, not just in your plans and preparations but also in your thinking. It involves recognizing that survival often depends on the ability to pivot and embrace new approaches in response to changing realities.

Community: Finally, the principle of community recognizes that survival is not a solitary endeavor. Building relationships with like-minded individuals, forming support networks, and collaborating on preparedness efforts can significantly enhance your resilience. The community offers a wealth of shared knowledge, resources, and mutual aid, making it a critical aspect of successful prepping. In times of crisis, a strong community can be the difference between thriving and merely surviving.

Together, these core principles form the foundation of the prepper lifestyle. They are interdependent, with each contributing to a holistic approach to preparedness that is both practical and philosophical. By understanding and integrating these principles into your preparedness plan, you set the stage for a life characterized by readiness, resilience, and self-sufficiency.

REDUNDANCY

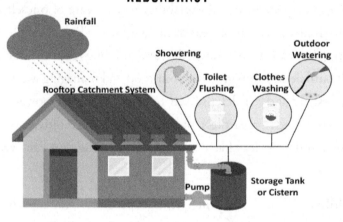

The concept of redundancy plays a pivotal role, serving as a safeguard against the unpredictable nature of crises and disasters. Redundancy, in its essence, refers to the strategic duplication of critical supplies, systems, and information to ensure that a single point of failure does not result in a total collapse of survivability. This principle is fundamental, for it embodies the prepper's adage: "Two is one, and one is none."

The Criticality of Backup Systems: Imagine a scenario where your primary water source becomes contaminated or inaccessible. Without a secondary source, your ability to hydrate, one of the basic necessities for survival, is severely compromised. Similarly, if your only means of generating power—be it solar panels or a generator—fails, you could be left in the dark, literally and figuratively. These are not mere hypotheticals; they are potential realities that underscore the importance of having backups. By establishing alternative water sources, such as rainwater collection systems or natural springs, and diversifying power generation methods, you significantly enhance your resilience against disruptions.

Redundancy in Food Supplies: Food security is another domain where redundancy is crucial. Depending solely on a single type of food storage—say, freeze-dried meals—exposes you to risk if that food source becomes unavailable or if dietary needs change. Incorporating variety, such as canned goods, home-preserved foods, and dry staples, alongside a rotation system, ensures nutritional needs are met even when external supply chains are disrupted. Furthermore, cultivating a garden provides a renewable source of fresh produce, adding another layer of redundancy to your food supply strategy.

Information and Skills Backup: Redundancy also applies to knowledge and skills. Relying on digital sources for critical information can be a vulnerability if power or internet access is lost. Keeping physical copies of important documents, manuals, and survival guides ensures access to vital information regardless of electronic availability. Similarly, cross-training within a community or family—ensuring multiple people know how to perform essential tasks like first aid, navigation, and mechanical repairs—guards against the loss of capabilities due to the absence or incapacity of a single individual.

Practical Application and Balance: While the principle of redundancy is paramount, it is equally important to strike a balance. Overstocking supplies without considering storage limitations or the practicality of maintaining multiple systems can lead to its own set of challenges. The key is to prioritize and apply redundancy to those areas most critical to survival and well-being, ensuring that you are prepared for the unexpected without overwhelming your resources and capacities.

Redundancy is not just about having more; it's about strategic preparation that anticipates and mitigates the risks of single points of failure. Through careful planning and thoughtful implementation of backup systems, supplies, and knowledge, preppers can create buffers against the uncertainties of the future, ensuring a higher degree of self-sufficiency and resilience in the face of adversities.

ADAPTABILITY

Adaptability, the capacity to adjust to new conditions and circumstances, stands as a cornerstone in the foundation of effective prepping. It's the recognition that no plan, however meticulously crafted, can account for every conceivable scenario. The true strength of a prepper lies not just in the depth of their supplies or the robustness of their shelter but in their ability to pivot and adapt when faced with the unpredictable.

The Role of Adaptability in Prepping: In the landscape of survival, conditions are in constant flux; what works today may not work tomorrow. A natural disaster, societal upheaval, or personal circumstance can alter the playing field, rendering prior plans less effective or even obsolete. Adaptability—modifying strategies, employing new tactics, and leveraging available resources to maintain or regain stability—is what allows a prepper to navigate these changes. It's about being resourceful, open-minded, and innovative in the face of challenges.

Cultivating an Adaptable Mindset: Fostering adaptability begins with mindset. Embrace the principle of lifelong learning. Be curious and seek out new skills, knowledge, and perspectives. This not only broadens your survival toolkit but also enhances your ability to think creatively about problem-solving. Encourage flexibility in your thinking by considering various scenarios and practicing how you would respond. Mental simulations, coupled with physical drills, can prepare you to act decisively and effectively under pressure.

Flexibility in Planning: When devising your prepping plans, incorporate flexibility. Build in contingencies and alternative options. For instance, if your primary evacuation route is blocked, having several pre-planned alternatives can mean the difference between safety and peril. Similarly, diversify your skills and supplies to avoid over-reliance on a single resource or strategy.

Learning from Experience: Adaptability also means being receptive to lessons learned from past experiences, both personal and observed. After any event or drill, conduct a thorough review of what worked, what didn't, and why. This reflective process is invaluable for identifying adjustments needed to improve future outcomes.

Building Resilience through Adaptability: Ultimately, adaptability enhances resilience, equipping you to face and overcome adversity. By staying mentally agile and open to change, you can adjust your strategies in real time, ensuring that you, your family, and your community are better prepared to withstand whatever the future may hold.

Adaptability in prepping is about expecting the unexpected and having the wherewithal to meet it head-on. Cultivating this quality ensures that when plans inevitably need to change, you can do so smoothly and effectively, maintaining your resilience and advancing your preparedness.

COMMUNITY INTEGRATION

The integration into and building of a community stands as a cornerstone in the edifice of preparedness and survival. In the journey of prepping, while individual resilience is vital, the role of community is unparalleled in enhancing survival chances. It transcends mere survival; it's about thriving in the face of adversity through the strength of collective effort and shared wisdom.

The Power of Collaboration: In the context of preparedness, the adage "Together, we stand; divided, we fall" could not be more apt. Collaboration within a community allows for the pooling of resources, skills, and knowledge, creating a synergy that magnifies the capabilities of individual's manifold. For instance, one member's expertise in gardening can complement another's proficiency in mechanical repairs, thereby ensuring that the community as a whole benefits from a diverse set of skills that might be critical in various scenarios.

Knowledge Sharing and Mutual Aid: The essence of community lies in the sharing of knowledge and mutual aid. Workshops, training sessions, and even informal gatherings provide opportunities for members to learn from each other, from basic survival skills to advanced medical training. This exchange of knowledge not only enhances the group's overall preparedness but also fosters a sense of belonging and responsibility toward each other. In times of need, this mutual aid becomes the safety net that can make the difference between despair and hope, providing emotional support alongside practical assistance.

Building Resilient Networks: The strength of a community is not just in its immediate numbers but also in its connections and networks. Establishing relationships with neighboring communities, local authorities, and emergency services can significantly enhance the resilience of a prepping community. These networks can facilitate the exchange of information, resources, and support on a larger scale, thereby improving the survival chances of all involved.

Cultivating a Culture of Preparedness: Beyond survival, a community can cultivate a culture of preparedness that extends to future generations. Through education and engagement, children and newcomers can be integrated into the prepping lifestyle, ensuring that the principles of self-sufficiency and mutual support are perpetuated.

Embracing Diversity for Strength: A community's strength also lies in its diversity. Diverse perspectives and experiences can lead to innovative solutions and strategies for dealing with challenges. Embracing this diversity and fostering an environment of inclusivity and respect is crucial for the cohesion and effectiveness of a prepping community.

The role of community in prepping cannot be overstated. By building and being part of a community that values collaboration, knowledge sharing, and mutual aid, individuals enhance not only their own chances of survival but also contribute to the resilience and well-being of the group as a whole. In the uncertain landscape of the future, such communities can serve as beacons of hope and bastions of strength, ready to face whatever challenges may come.

LESSONS FROM THE FIELD

COMMON MISTAKES IN PREPPING

Prepping is a complex and nuanced journey, one that requires careful planning, foresight, and adaptability. However, even the most diligent preppers can fall prey to common mistakes that can undermine their preparedness efforts. By examining real-world examples and hypothetical scenarios, we can illuminate these pitfalls and learn how to avoid them.

Underestimating Resource Needs: A classic oversight in the prepping community is the underestimation of resource needs, particularly for water and food. For instance, a family of four may stockpile what they believe to be a three-month supply of water, only to find out during a prolonged power outage that their actual consumption far exceeds their estimates. This miscalculation can stem from not accounting for additional needs, such as hygiene or cooking. This highlights the importance of realistic usage assessments and regular testing of your preparedness plan.

Overreliance on a Single Plan: Another frequent mistake is putting all of one's eggs into one basket, so to speak. A prepper might design an elaborate bug-out plan that relies heavily on a single route and mode of transportation. However, if an unexpected natural disaster renders that route impassable or the vehicle unavailable, the lack of an alternate plan can lead to chaos. This scenario underscores the necessity of having multiple, flexible plans that can adapt to the unexpected.

Neglecting Security and Self-Defense: In the realm of prepping, there's often an enthusiastic focus on food and water storage, while security and self-defense are sometimes overlooked. Even if preppers meticulously stockpile supplies, it may do no good if they fail to secure their storage locations adequately. In the event of a break-in, they may lose a significant portion of their provisions. It is important to consider all aspects of safety and security in your prepping strategy.

Ignoring Mental Health and Morale: The psychological aspect of survival is frequently underestimated. The isolation and lifestyle changes of long-term, off-grid living can cause mental strain. The lack of preparation for maintaining mental health and morale can lead to interpersonal conflicts and a breakdown in group cohesion. Mental health is just as crucial as physical preparedness.

Failing to Practice Skills Regularly: Owning survival gear and knowing how to use it in theory is one thing; being proficient in its use is another. A prepper who purchases advanced medical kits but never practices using them can find themselves overwhelmed when a real emergency occurs. Regular practice and skill development ensures that when the time comes, you're not just equipped but also capable.

By recognizing and learning from these common mistakes, preppers can refine their strategies, making them more comprehensive, realistic, and effective. The key lies in constant evaluation, adaptation, and the willingness to learn from both personal experiences and the experiences of others.

ANALYTICAL BREAKDOWN

Learning from mistakes—whether one's own or those of others—is invaluable. Here, we'll dissect a few common pitfalls that preppers encounter, providing an analytical breakdown to understand what went wrong, why it happened, and the consequences. This examination serves as a roadmap for avoiding similar missteps in your prepping journey.

Scenario 1: Underestimating Water Needs

The Mistake: A family prepared for a short-term natural disaster by stockpiling what they believed to be an adequate water supply. They based their calculations on minimal daily usage, not accounting for cooking, cleaning, or the possibility of extended recovery times.

Analysis: The critical error was the underestimation of water needs. The family calculated their requirements solely on direct consumption, neglecting other uses of water that become critical over time. This oversight stems from a common misconception that survival priorities are static and unchanging.

Consequences: The family quickly exhausted their water supply, leading to dehydration and the risk of waterborne diseases from unsanitary conditions. The situation forced them into a desperate search for additional sources under hazardous conditions.

Scenario 2: Overreliance on a Single Plan

The Mistake: A prepper meticulously planned for an economic collapse by investing all their resources into gold and silver, neglecting other forms of wealth preservation and practical supplies.

Analysis: This strategy faltered due to its inflexibility and lack of diversification. By focusing solely on one form of asset protection, the prepper failed to consider scenarios where liquidity or access to markets could be limited, or where basic supplies would become more immediately valuable than precious metals.

Consequences: When a localized disaster struck, rendering markets and banking services inoperative, the prepper was unable to trade or utilize their investments for essential goods and services, highlighting the flaw in depending on a single plan for all eventualities.

Scenario 3: Neglecting Community and Network Building

The Mistake: An individual prepper chose to prepare in isolation, believing that self-sufficiency meant going it alone and avoiding any reliance on external support.

Analysis: This approach overlooked the significant benefits of community and network building. In times of crisis, individuals often require support beyond their capabilities, whether through shared knowledge, resources, or mutual aid. The mistake lay in equating self-reliance with isolation, ignoring the strength found in collective resilience.

Consequences: When faced with a challenge that exceeded their personal resources, the prepper struggled without the support network that could have offered assistance. This isolation not only compounded the physical difficulties but also the psychological strain of facing adversity alone.

These scenarios underscore the complexity of preparedness and the necessity of a holistic approach. By recognizing and learning from these common mistakes, preppers can refine their strategies, ensuring a more resilient and adaptable posture against the myriad challenges that may arise.

PROACTIVE GUIDANCE: AVOIDING PITFALLS

To navigate the complex landscape of prepping effectively, it's essential to learn from past errors and implement strategies that mitigate risks. Here are actionable tips and techniques designed to guide you away from common prepping pitfalls, ensuring a more comprehensive and resilient approach to preparedness.

Comprehensive Resource Planning: To avoid underestimating your resource needs, especially water, adopt a holistic planning approach. Consider all possible uses for resources in various scenarios, including hygiene, cooking, and extended emergencies. The Federal Emergency Management Agency (FEMA) recommends storing at least one gallon of water per person per day for at least three days for drinking and sanitation (FEMA, 2021). Diversifying your storage solutions and regularly revisiting your plan to adjust for changes in household size or needs can ensure you're better prepared.

Diversification of Plans and Supplies: Overreliance on a single plan or asset, such as precious metals, can be risky. Diversify your investments and supplies to include a range of assets and essentials that will be valuable in various scenarios. Incorporate cash, tradeable goods, and practical survival items into your strategy. Maintaining a balance between tangible assets and practical supplies ensures you have the flexibility to respond to different challenges.

Building and Maintaining Community Ties: The significance of community cannot be overstated. Engage with local prepping groups, participate in community resilience projects, and contribute to mutual aid networks. These connections can provide not just logistical support but also invaluable knowledge exchange and emotional sustenance in times of crisis. Start by attending local meetings or workshops and gradually build trust and collaborative relationships.

Continuous Learning and Adaptability: Adopt a mindset of continuous learning and openness to new information. Regularly update your knowledge base through reputable sources, training, and practice. This proactive approach to learning enhances your adaptability, allowing you to adjust plans as new information or technologies become available.

Psychological Preparedness: Cultivate mental resilience through stress management techniques, scenario planning, and maintaining a positive outlook. Psychological preparedness is as crucial as physical supplies in navigating the challenges of emergencies effectively.

Steering clear of common prepping pitfalls requires a multifaceted approach that encompasses thorough planning, diversification, community engagement, continuous learning, and psychological resilience. By integrating these strategies into your prepping practices, you can build a more robust and adaptable preparedness plan capable of withstanding the uncertainties of the future.

As we close this chapter on "Essential Prepping Strategies," it's crucial not just to absorb these insights but to integrate them actively into your preparedness plans. The journey of a prepper is both challenging and rewarding, demanding constant vigilance, learning, and adaptation. The strategies discussed herein—from ensuring redundancy in your supplies to building a resilient community—form the bedrock upon which you can construct a comprehensive and robust approach to prepping. Remember, preparedness is not a destination but a continuous journey.

Now, armed with a deeper understanding of common pitfalls and how to avoid them, it's time to turn knowledge into action. Assess your current prepping strategies, identify areas for improvement, and commit to making the necessary adjustments. The path ahead requires diligence, commitment, and the willingness to evolve.

Thank you for purchasing The Complete Prepper's Survival Bible. Scan the QR code above with your smartphone for access to 10 exclusive bonuses. Alternatively, you can visit this website: https://rebrand.ly/scan-for-your-bonuses

PART 2:
PRACTICAL PREPPING SKILLS

CHAPTER 4
ADVANCED PREPPER PANTRY MANAGEMENT

"By failing to prepare, you are preparing to fail."
— ***Benjamin Franklin***

The words of Benjamin Franklin ring true, especially when we turn our focus to the heart of home preparedness: the prepper pantry. A well-managed pantry is more than just a place to store food; it's the cornerstone of long-term survival and self-sufficiency. Ensuring you have a reliable food source, regardless of what life throws your way, is not just smart—it's essential.

In this chapter, we dive into the critical components that make up an advanced prepper pantry. Designing a pantry that meets your specific needs can significantly enhance your ability to maintain a sustainable food source. We'll explore how to craft a space that not only stores food but does so in a way that keeps it accessible and organized. This includes considering the layout of shelves, the best materials to use, and how to maintain an optimal environment for food preservation.

Pest management is another critical aspect we'll cover. The last thing you want is for your hard-earned supplies to be compromised. We'll provide you with strategies to prevent infestations and, if necessary, how to deal with them safely and effectively.

Finally, we'll delve into storage solutions that go beyond traditional shelving. In the world of prepping, creativity is key. From utilizing hidden spaces to implementing vertical gardens, these innovative ideas can significantly expand your storage capabilities and ensure your pantry is prepared for any scenario.

By the end of this chapter, you'll have a solid foundation on how to build and maintain a prepper pantry that not only serves your needs today but secures your food independence for the future. Let's get started.

PANTRY DESIGN

The blueprint of your pantry is more than just shelves and doors—it's a master plan for ensuring your survival and comfort through any situation. An optimally designed pantry transcends mere storage; it becomes a dynamic space where every item has a purpose and every square inch is utilized to its fullest potential. Let's unlock the secrets to creating a pantry that not only stores food but does so in a manner that maximizes longevity, accessibility, and efficiency.

DESIGN FUNDAMENTALS

The foundation of a well-organized pantry is not just about stacking cans and bags. It's about creating a system where every item is easy to find, and every space serves a purpose. An efficient layout, strategic organization, and thoughtful accessibility are paramount. Here's how you can achieve that:

- **Layout:** Start with a clear plan of where everything will go. Designate areas for canned goods, dry foods, spices, cooking supplies, and equipment. This segmentation helps in locating items quickly and managing inventory effectively.

- **Organization:** Utilize shelving units that match the size and type of items you're storing. Adjustable shelves offer flexibility as your storage needs change. Consider using clear, labeled bins for smaller items to prevent them from getting lost in the shuffle.

- **Accessibility:** Ensure that frequently used items are within easy reach. Place lesser-used items higher up or in less accessible areas. Incorporating pull-out drawers or lazy Susans for corners can make use of hard-to-reach spaces.

- **Light and Temperature Control:** Protect your supplies from spoilage by keeping your pantry cool, dry, and away from direct sunlight. LED lights can provide excellent visibility without raising the temperature.

- **Space Utilization:** Make the most of your pantry's vertical space with floor-to-ceiling shelving. Over-the-door racks and hooks can hold tools and smaller items, maximizing every available inch.

Incorporating these fundamentals into your pantry design ensures a robust system that stands up to the demands of prepping. Whether you're facing a short-term power outage or a long-term supply chain disruption, your pantry will be your family's lifeline, reliably supporting your nutritional needs and reducing the stress of uncertainty.

CUSTOMIZATION STRATEGIES

Crafting a pantry that mirrors your unique needs and preferences isn't just beneficial—it's essential. A one-size-fits-all approach simply doesn't cut it when it comes to prepping. Each household has unique rhythms, dietary preferences, and constraints. Customizing your pantry is about weaving your personal story into the very fabric of your preparedness strategy. It's about creating a space that not only stores food but also aligns with your lifestyle, ensuring that in times of need, your pantry feels like a continuation of your home, not just a survival bunker.

Personalized Pantry Design

The journey to a personalized pantry begins with a clear understanding of your specific needs. Space, budget, and dietary restrictions play significant roles in shaping your pantry design. Here's how to tailor your pantry to fit your unique scenario:

- **Space Considerations:** Not everyone has the luxury of a large pantry. For those with limited space, think vertically and use wall-mounted shelves and hanging organizers. If you're in a tiny apartment, consider under-bed storage or repurpose a closet into your pantry area. It's about making the space work for you.

- **Budget-Friendly Solutions:** Customization doesn't have to break the bank. Use recycled containers for storage and scour thrift shops for shelving units and baskets. Creativity in finding and repurposing items can lead to a highly functional pantry without a hefty price tag.

- **Dietary Adaptations:** Tailor your pantry to your dietary needs by allocating specific zones. If you have gluten-free members, designate an area for their foods. For vegetarians or those with food allergies, clear labeling and separate storage can prevent cross-contamination and make meal preparation easier.

- **Guiding Questions:** To ensure your pantry truly reflects your needs, ask yourself:
  o What foods do we eat regularly, and how can I ensure they're well-stocked and accessible?
  o How can I make the most of my available space to store the essentials?
  o What storage solutions fit within my budget while still meeting my organizational needs?

By addressing these questions, you can create a pantry that not only meets your storage needs but also your lifestyle preferences. Whether you're dealing with dietary restrictions, budget constraints, or space limitations, there are countless ways to customize your pantry. The goal is to design a space that supports your prepping efforts seamlessly, ensuring you and your loved ones can face any challenge with confidence.

PEST MANAGEMENT

A fortress is only as strong as its defenses, and in the realm of food storage, the battle against pests is ongoing. Effective pest management in a prepper's pantry isn't merely a reactive measure; it's a proactive strategy that begins with prevention. Keeping pests at bay ensures the integrity and longevity of your supplies, safeguarding your peace of mind along with your provisions. From the foundational cleanliness to the strategic use of storage solutions, integrating pest deterrent measures into your pantry design is a critical step in maintaining a secure and reliable food supply.

PROACTIVE PREVENTION TECHNIQUES

Cleanliness: The first line of defense against pests is a clean pantry. Regularly sweep and vacuum to remove crumbs and residues that attract pests. Wipe down shelves with a vinegar solution to discourage ants and other insects.

Airtight Containers: Store grains, flours, dried fruits, and nuts in airtight containers. Not only do these containers keep food fresh, but they also prevent pests from detecting and accessing your food supplies.

Natural Deterrents: Certain natural substances can deter pests without the need for harsh chemicals. Bay leaves, for example, can be placed in containers of grains and flours to repel weevils. Mint, lavender, and cedar blocks can ward off a variety of pests and can be integrated into the pantry design for a seamless defense.

Proper Food Rotation: Regular rotation of your food supplies minimizes the chance of infestation by ensuring that older items are used first, reducing the time pests have to discover and infest items.

Sealed Entry Points: Inspect your pantry and the surrounding area for any cracks, holes, or gaps where pests might enter. Seal these entry points with caulk or steel wool, especially around pipes and wires that enter walls.

By adopting these preventative measures, you're not just protecting your food; you're preserving the functionality and effectiveness of your entire prepping strategy. A pantry designed with pest prevention in mind is a cornerstone of a secure, sustainable food storage system, ensuring that when you reach for your supplies, they are as ready and resilient as you are.

REMEDIATION METHODS

Even the most diligently managed pantries can sometimes fall prey to the tenacity of pests. Recognizing the inevitability of these unwelcome guests is the first step toward maintaining the sanctity of your provisions. The challenge, then, isn't just about prevention; it's about being prepared to act swiftly and effectively when infestations arise. The key is to employ remediation methods that are safe, effective, and sustainable, ensuring that your pantry remains a stronghold of nourishment, not a haven for pests.

EFFECTIVE PEST REMOVAL STRATEGIES

When pests breach your defenses, a quick and decisive response is crucial. Here are several strategies for reclaiming your pantry without resorting to harsh chemicals that could compromise your food's safety and your health:

- **DIY Natural Remedies:** Many household items can be transformed into effective pest deterrents. For example, a mixture of water and vinegar can deter ants, while diatomaceous earth is an excellent all-around pest deterrent that's safe to use around food. Herbs like bay leaves and mint can also repel insects, adding a fragrant layer of protection to your pantry.

- **Advanced Technological Solutions:** For more persistent infestations, technology offers non-toxic solutions. Ultrasonic pest repellers emit frequencies that are intolerable to pests but harmless to humans and pets. These devices can be an unobtrusive and continuous line of defense against invaders.

- **Physical Traps:** Sometimes, the old ways are the best ways. Traps, whether for rodents or insects, can be an effective and immediate solution. Opt for traps that are safe to use in proximity to food and that minimize suffering for the creatures they catch.

A word of caution: while chemical solutions might offer a quick fix, their use in a food storage environment should be a last resort. The potential for contamination, not to mention the health risks to humans and pets, makes them less desirable. Whenever possible, prioritize methods that are safe and sustainable, preserving the integrity of your pantry and the well-being of your household.

Through a combination of vigilance and smart, safe remediation strategies, you can protect your prepper pantry from the inevitable intrusion of pests. The goal is not just to react but to do so in a way that upholds the principles of safety, sustainability, and self-sufficiency.

STORAGE SOLUTIONS

In the world of prepping, space is not just a luxury; it's a resource. Efficient use of available space is paramount for creating a pantry that not only stores ample supplies but does so in a way that keeps everything accessible and orderly. The key to maximizing pantry space lies not just in how much you can fit into it, but in how intelligently you can organize and access what you have. Through innovative storage solutions and strategic organization, even the most limited spaces can be transformed into a haven of readiness and efficiency.

SPACE OPTIMIZATION TECHNIQUES

Leveraging Vertical Space:

- **Ceiling-Mounted Shelves:** Often overlooked, the space just below the ceiling is ideal for items used infrequently. This could include bulk supplies or emergency kits that need to be out of the way but still within reach when needed.
- **Under-Shelf Baskets:** These baskets attach to existing shelves, effectively creating an additional storage layer. They're perfect for smaller items that can get lost in larger spaces, such as spice jars or packets of seeds.

Making Use of Doors:

- **Door Racks:** The back of the pantry door can serve as an extra wall of storage. Adjustable racks can hold everything from cans and jars to over-the-door organizers for smaller items.
- **Hanging Organizers:** Similar to those used for shoes or accessories, hanging organizers can be repurposed in the pantry to store a variety of goods, ensuring that every square inch of the door space is put to good use.

Case Study: A small household transformed its tiny pantry using these techniques, doubling its storage capacity. Ceiling-mounted shelves were installed to store large, lightweight items, such as paper towels and bulk cereals, while under-shelf baskets held kitchen towels and utensils. The back of the door was outfitted with multiple racks, creating a system that sorted items by type and frequency of use. This approach not only maximized the available space but also made inventory management and access significantly more straightforward.

CREATIVE STORAGE

When it comes to prepping, traditional storage methods can sometimes fall short of meeting our comprehensive needs. This is where innovation steps in, transforming every nook and cranny into potential storage goldmines. Embracing creative storage solutions not only maximizes your space but also infuses your pantry management with a sense of ingenuity and adaptability. Let's explore how unconventional approaches can revolutionize the way you store your essentials, making your pantry not just a place for food, but a masterpiece of efficient design.

INGENIOUS STORAGE SOLUTIONS

The art of storage is not just about putting things away but doing so in a way that enhances accessibility, preserves quality, and optimizes space. Here are some creative strategies that push the boundaries of traditional storage:

- **Vertical Gardening:** This innovative approach isn't just for fresh produce. Use vertical spaces to grow herbs and small vegetables. Incorporating a living pantry element adds freshness and vitality. Wall-mounted planters or hanging gardens can transform unused vertical spaces into lush, productive areas.

- **Hidden Compartments:** The concept of hidden storage is not new, but its application in pantry design can be both fun and functional. Consider false bottoms in drawers, toe-kick drawers at the bottom of cabinetry, or even a hidden pantry behind what appears to be a cabinet facade. These elements not only provide extra storage space but also keep your pantry organized and your items out of sight, preserving a clean and uncluttered appearance.

- **Convertible Furniture:** Furniture with built-in storage or that can transform into storage units offers a dual-purpose solution for tight spaces. A table that doubles as a storage container or benches with lift-up seats can hide away bulk items or lesser-used appliances, keeping them accessible but out of the way.

As this chapter closes, let the journey to a more resilient and prepared future begin. With the insights and strategies shared, you possess the blueprint to transform your pantry into a stronghold of sustainability and security. Embrace the call to refine your space, safeguard against pests, and optimize every square inch with purpose and precision. Your pantry is not just a storage area—it's the heart of your home's preparedness, a testament to your commitment to safeguarding your well-being no matter what challenges lie ahead.

Now, with your pantry set for efficiency and resilience, what's next on the path to self-sufficiency? The upcoming chapter ventures into the vital realm of water purification and storage—a cornerstone of survival often underestimated until it's critically needed. How can you ensure that your water supply is as secure and reliable as your food stockpile? Prepare to dive deep into strategies that guarantee clean, accessible water, ensuring your preparedness plan is watertight in every sense.

Elevate your pantry prowess with "The Ultimate Prepper's Pantry Cookbook." It's one of 10 bonuses you'll receive to help you in your prepping journey. Simply scan the QR code above with your smartphone. Alternatively, you can visit this website: https://rebrand.ly/scan-for-your-bonuses

CHAPTER 5
MASTERING FOOD PRESERVATION AND STOCKPILING

"Preserving food is a celebration of nature's bounty and a testament
to our resilience in the face of uncertainty."

— Michael Pollan

Food preservation and stockpiling are not just practices; they're arts—arts that secure independence and readiness in the face of uncertainty. This chapter explores a tapestry of techniques and strategies designed to fortify your food security through thick and thin. Delving into the realms of advanced canning, strategic selection, and the wise rotation of supplies, we chart a course toward a pantry that stands as a bulwark against the unforeseen.

We'll uncover the nuances of pressure canning, vacuum sealing, and fermenting—methods that lock in nutrition and taste, extending the bounty of today into the security of tomorrow. With precision, we'll guide you through the intricacies of these processes, ensuring that safety and efficiency are paramount in your efforts to preserve.

Moreover, the selection of food for your pantry is anything but arbitrary. It demands a strategic approach that weighs longevity, versatility, and nutritional value, crafting a stockpile that supports both physical health and morale over the long haul.

But what of the foods once they're stored? Here, the principle of rotation comes to the fore, guarding against waste and ensuring that each meal from your pantry is as fresh and nutritious as possible. Coupled with savvy inventory management, this system keeps your provisions in perpetual readiness.

As we prepare to turn the page, remember: the practice of food preservation and stockpiling is one of empowerment. It's a declaration of independence from the whims of circumstance, a testament to the resilience and foresight that define the prepared. With the insights and instructions laid out in this chapter, your path to mastering these essential skills is clear. Are you ready to take the first step?

ADVANCED CANNING TECHNIQUES

In the quest for long-term food security, canning emerges as a linchpin, enabling us to preserve the bounty of today for the uncertainty of tomorrow. This method isn't just about survival; it's a bridge to maintaining a nutritious and varied diet when fresh food may not be an option. Canning, with its roots deep in history, has evolved with technology, offering us advanced techniques, like pressure canning, vacuum sealing, and fermenting. Each method comes with its intricacies and rewards, ensuring that the food preserved is as close to its original state in terms of nutrition and taste.

COMPREHENSIVE INSTRUCTIONS

Pressure Canning: Pressure canning is essential for preserving low-acid foods, like vegetables, meats, and fish, safeguarding against botulism. The process involves:

1. **Preparing the Food:** Start by selecting fresh, high-quality ingredients. Wash, peel, and cut as required.
2. **Filling the Jars:** Leave appropriate headspace as indicated for each food type. Remove air bubbles by gently tapping the jar.
3. **Sealing:** Wipe the jar rims, place the lids on them, and tighten the bands without over-tightening.

4. **Processing:** Place the jars in the pressure canner, lock the lid, and heat. Once steam flows steadily from the vent, close it and start counting the processing time as the pressure reaches the specified level.

5. **Cooling:** After processing, turn off the heat, let the canner cool naturally, and wait for the pressure to drop before opening. Remove the jars, allow them to cool for 12-24 hours, check their seals, label them, and store them.

Vacuum Sealing: Ideal for dry foods or those you'll freeze, vacuum sealing extends shelf life by removing air:

1. **Prepare Foods:** Dry or freeze wet items, like meat or cooked foods, to prevent liquid from interfering with the seal.

2. **Bagging:** Place the food in vacuum seal bags, leaving space at the top.

3. **Sealing:** Insert the open end of the bag into the vacuum sealer. The machine removes the air and seals the bag.

4. **Storage:** Label the bags with the contents and date. Store in a cool, dark place or freeze, if applicable.

Fermenting: Fermentation enhances food with probiotics, flavors, and longevity. Basic steps include:

1. **Preparation:** Choose fresh, organic produce. Cut or shred vegetables as desired.

2. **Brining:** Dissolve salt in water to create a brine. The salt concentration will depend on the recipe.

3. **Packing:** Place the prepared food tightly in jars, leaving adequate space from the top. Pour brine over, ensuring the food is completely submerged.

4. **Fermenting:** Seal the jar loosely to allow gases to escape or use a water seal. Store at room temperature, away from direct sunlight. Fermentation time varies by recipe.

5. **Storing:** Once fermentation is complete, tighten the lids and store the jars in a cool, dark place or refrigerate.

By applying these techniques you can turn your pantry into a repository of healthy, ready-to-use foods.

SAFETY PROTOCOLS

Navigating the world of canning requires not just skill and patience but a steadfast commitment to safety. The preservation of food, while a gateway to self-sufficiency, carries the weight of responsibility to prevent foodborne illnesses. Botulism, a rare but serious illness, casts a long shadow over improper canning practices, reminding us of the critical need for vigilance and adherence to safety guidelines.

GUARDING AGAINST BOTULISM AND OTHER HAZARDS

Understanding Botulism: Botulism, caused by the bacterium Clostridium botulinum, thrives in low-oxygen environments, making improperly canned foods a potential risk. The toxin produced by these bacteria can lead to severe health consequences, emphasizing the importance of meticulous canning practices.

Key Safety Measures:

1. **Proper Equipment:** Always use canning jars and lids designed for the task. Regular glass jars or reused commercial jars may not withstand the pressures of canning, leading to breaks or improper sealing.
2. **Sterilization:** Before filling, sterilize jars and lids by boiling them for at least 10 minutes. This step is crucial in killing any pre-existing bacteria.
3. **Use the Right Method:** Low-acid foods (like vegetables, meats, and fish) must be processed using a pressure canner to achieve temperatures high enough to kill botulinum spores. High-acid foods (such as fruits and pickles) can be safely canned using a water bath method.

4. **Follow Trusted Recipes:** Use recipes from reputable sources that include specific canning times and pressures. These guidelines are designed to ensure safety and minimize the risk of contamination.

5. **Inspect Your Cans:** Before consumption, inspect cans for signs of spoilage, such as bulging lids, leaks, or off-odors. When in doubt, throw it out.

Real-World Consequences: A case that underscores the importance of adhering to safety protocols involved a family reunion where home-canned beans were served. The beans, canned using incorrect methods, led to multiple cases of botulism. This incident, among others, highlights the crucial nature of safety in canning. It serves as a stark reminder that the joy of sharing home-preserved foods comes with the duty to ensure they are safe to eat.

Safety in canning is not just a guideline; it is the foundation upon which the entire practice rests. By following these protocols, you protect not just the integrity of your food but the health and well-being of those who consume it. As we delve further into the intricacies of food preservation, let the principle of safety guide every jar you seal, ensuring your pantry is a source of nourishment, not risk.

STRATEGIC FOOD SELECTION

The art of stockpiling is not in the accumulation but in the careful selection of what to store. As we turn our attention to the strategic selection of food for long-term storage, we're not just filling shelves; we're curating a collection that ensures longevity, versatility, and nutritional balance. The choices we make today lay the groundwork for a robust pantry that stands ready to support us, come what may.

SELECTION CRITERIA

Longevity:

- **Grains and Legumes:** Rice, beans, lentils, and whole grains offer a solid foundation, boasting shelf lives that extend to years when stored properly. These staples not only provide essential carbohydrates and proteins but also serve as the backbone of countless meals.
- **Dried Pasta:** With a shelf life that rivals that of grains, dried pasta is a versatile staple that can be dressed up in numerous ways, ensuring mealtime variety.

Versatility:

- **Canned Vegetables and Fruits:** These can be used straight from the can or incorporated into more complex dishes. Opt for a variety of colors and types to ensure a wide range of vitamins and minerals.
- **Powdered Milk and Eggs:** These provide the functionality of their fresh counterparts but with the convenience of long-term storage. These are ideal for baking or as protein sources.

Nutritional Value:

- **Nuts and Seeds:** High in healthy fats, protein, and various micronutrients, nuts and seeds are not only nutritious but also have a relatively long shelf life when stored in airtight containers.

- **Freeze-dried Meats and Vegetables:** While more expensive, they offer the closest nutritional profile to fresh options and rehydrate quickly for use in meals.

Ease of Use:

- **Instant Meals and Soups:** For situations where cooking is either not possible or practical, having a stock of ready-to-eat meals can be invaluable. These should be used sparingly and complemented with more nutritionally dense options.
- **Spices and Seasonings:** Often overlooked, a good stock of various spices and seasonings can transform basic ingredients into delicious meals, encouraging the consumption of stored food even outside of emergency scenarios.

Organizing Your Selection:

- A simple table or bullet list categorizing foods by type (Grains, Proteins, Vegetables/Fruits, etc.), along with their expected shelf life and nutritional highlights, can serve as a quick reference for both shopping and rotating stock.

By embracing these criteria in your selection process, you can ensure your pantry is not just a food store but a well-rounded resource. This strategic approach to food selection not only maximizes the utility and shelf life of your stockpile but also ensures that, regardless of what challenges you may face, your nutritional needs are met with a diverse and balanced diet.

PREPARATION AND VARIETY

The secret to a resilient pantry lies not just in its capacity but in its diversity. Variety in food selection combats the monotony of palate fatigue and is crucial in ensuring a balanced diet. This variety also plays a pivotal role in meal preparation, offering flexibility that can cater to any situation, be it a quick lunch or a more elaborate dinner that brings a sense of normalcy in times of stress.

EMBRACING DIVERSITY IN YOUR PANTRY

The Importance of Meal Variability:

Imagine opening your pantry to find not just food but a world of culinary possibilities. From the simplicity of rice and beans to the comforting warmth of a vegetable stew, the key to a satisfying meal plan lies in the variety of your stockpile. Each ingredient should serve multiple purposes, allowing for a range of meals that can keep both body and spirit nourished.

Easy Preparation Meets Nutritional Needs:

- **Ready-to-eat Meals:** In situations where time or resources are limited, having meals that require minimal preparation but still offer nutritional value is a game-changer.
- **Bulk Cooking Ingredients:** Items like quinoa, which can serve as the base for salads, soups, and side dishes, are invaluable. They're not only easy to prepare but can be flavored in countless ways to suit any taste.

Incorporating Global Flavors:

- Broadening your culinary horizons can transform your meal preparation from a chore into an adventure. Stocking spices from around the world enables you to explore international cuisines right from your kitchen. Consider, for instance, creating a Moroccan tagine using preserved lemons and olives or an Italian pasta dish with sun-dried tomatoes and canned artichokes.
- This approach not only enriches your diet with a spectrum of flavors but also introduces a variety of nutrients from a wide range of ingredients, ensuring a balanced diet.

Meal Examples from Stored Foods:

- **Breakfast:** Oats with freeze-dried berries, nuts, and powdered milk.
- **Lunch:** Quinoa salad with canned chickpeas, dried fruits, and nuts.
- **Dinner:** Pasta with vacuum-sealed vegetables and a sauce made from canned tomatoes and spices.

By embracing variety and simplicity in meal preparation, your pantry becomes more than a food store—it becomes a treasure trove of flavors and nutrients, ready to meet your needs whatever the circumstance. This strategic approach to food selection ensures that your meals remain interesting, enjoyable, and, most importantly, nutritious, turning each dining experience into an opportunity for exploration and satisfaction.

SUSTAINABLE ROTATION SYSTEM

The essence of a well-maintained pantry lies in its ability to evolve and adapt—not just a static collection of items but a dynamic, rotating system that ensures freshness and nutrition. The implementation of a sustainable rotation system transforms the concept of stockpiling from mere accumulation to an art form, where every item is utilized at its peak, minimizing waste and maximizing benefit.

ROTATION PRINCIPLES

The Core of Rotation: At the heart of an effective pantry is the "first in, first out" (FIFO) principle. This simple yet profound strategy ensures that foods are consumed in the order they are added to the pantry, keeping the stock fresh and reducing the risk of spoilage.

Applying FIFO in the Home Setting:

- **Label Clearly:** Every item in your pantry should be clearly labeled with its purchase or canning date. This simple step makes it easier to identify and use older items before newer ones.

- **Organize Strategically:** Design your storage spaces to facilitate easy access to older items, perhaps by placing them in front or using a designated shelf for items that need to be used soon.

- **Categorize by Type and Date:** Keep similar items together and organize them by date. For example, all canned vegetables should be in one area, with the oldest cans in front.

- **Regular Reviews:** Make it a habit to periodically review your pantry's contents. This not only helps in rotating stock but also in planning meals around items that need to be consumed.

Adapting to Different Foods and Conditions:

- Not all foods have the same shelf life or storage requirements. For instance, grains and dried beans are more forgiving than oils and nuts, which can go rancid. Adjust your rotation strategy to account for these differences, prioritizing the rotation of more perishable items.

- Consider environmental factors, such as temperature and humidity, which can affect food quality. Items stored in less-than-ideal conditions may need to be rotated more frequently.

By integrating these rotation principles into your pantry management, you can ensure that your stockpile remains vibrant and nutritious. This active approach to food storage not only safeguards your investment but also enriches your daily meals, making the most of every item you've taken care to store.

INVENTORY MANAGEMENT

In the world of food preservation and stockpiling, having a bountiful supply is only half the battle. The other, equally crucial half is knowing what you have, where it is, and how long it will last. This is where effective inventory management becomes indispensable, transforming a mere collection of food items into a well-organized, accessible, and up-to-date stockpile. Whether through digital means or traditional manual tracking, maintaining an accurate inventory ensures that no item is forgotten, and every part of your stockpile contributes to your food security.

NAVIGATING INVENTORY OPTIONS

Digital Tracking Systems: In the digital age, inventory management can be as convenient as tapping on a screen. Numerous apps and software are designed to help you track your stockpile, offering features like:

- **Expiration Alerts:** Notifications to use items before they spoil.

- **Quantity Tracking:** Keep tabs on how much of each item you have, making shopping more efficient.
- **Nutritional Information:** Some apps allow you to log nutritional data, helping you plan balanced meals.

Digital tools provide an all-in-one solution for the modern prepper, combining convenience with comprehensive stockpile oversight.

Manual Methods: For those who prefer a more hands-on approach or seek simplicity, manual inventory methods have stood the test of time:

- **Inventory Notebook:** A dedicated notebook where each page represents a category of supplies, updated by hand as items are added or used.
- **Printed Sheets:** Similar to the notebook but allows for sheets to be replaced or reordered as needed, keeping the inventory flexible and current.

Choosing Your Method: The choice between digital and manual inventory methods depends on your personal preferences, lifestyle, and how interactively you wish to manage your stockpile. While digital systems offer ease and efficiency, manual methods provide a tactile and straightforward approach that many find satisfying and reliable.

Effective inventory management, regardless of the method, ensures that your stockpile remains a dynamic resource that adapts to your needs, preventing waste and ensuring that when the time comes, your pantry is not just stocked but stocked wisely and ready to support you.

Safeguarding your family's food supply is critical. Reinforce your prepping and survival skills by scanning the QR code above with your smartphone. You'll gain access to 10 bonuses ranging from food preservation, to seed saving, to ethical hunting, and more.
Alternatively, you can visit this website:
https://rebrand.ly/scan-for-your-bonuses

CHAPTER 6
WATER SECURITY AND PURIFICATION

"Water is life's matter and matrix, mother and medium. There is no
life without water."

— *Albert Szent-Györgyi*

Water, transparent, tasteless, odorless, and nearly colorless, is the
cornerstone of survival. This chapter unfolds the essence of securing and
purifying this vital resource, ensuring you're prepared for any scenario. In the
face of emergencies, understanding how to store, purify, and collect water is
not just about survival; it's about maintaining a semblance of normalcy in
chaotic times. Here, we'll dive into strategies for comprehensive water

storage, introduce you to cutting-edge purification techniques, and guide you on becoming self-reliant in water collection and purification.

Imagine a situation where taps run dry and stores are out of bottled water. What would you do? This isn't a call to panic but to prepare. By the end of this chapter, you'll be equipped with the knowledge to secure a safe water supply for yourself and your loved ones, come what may. From the simplest methods to gather and store water to employing advanced technologies for purification, we'll cover all bases. Plus, we'll explore the independence that comes from knowing how to harvest rainwater or purify a murky pond into drinking water.

The journey to water security starts with a plan. How much water do you need? Where will you store it? How can you ensure it stays clean? We'll tackle these questions head-on, providing clear, actionable advice. Whether you live in an apartment with limited space or a house with a yard, you'll discover storage solutions that work for you.

But what if your stored supply runs low? We've got you covered with innovative and straightforward purification methods that can turn suspect water into a lifeline. From the convenience of UV light purifiers to the simplicity of solar disinfection, we'll cover it all.

And because self-reliance is key, we'll also guide you through setting up a rainwater collection system and tapping into natural water sources safely. With this knowledge, you can not only secure your water needs but also take a significant step towards a sustainable lifestyle.

Let this be your call to action: assess your current water security strategy and take steps to bolster it. The empowerment and peace of mind that come from ensuring access to clean, safe water are unmatched.

COMPREHENSIVE WATER STORAGE

In the landscape of emergency preparedness, securing a reliable water supply stands paramount. It's not just about having water; it's about having enough of it, stored correctly, to meet every need—from the sip that quenches your thirst to the water that cleanses your hands. The guidance from the Centers for Disease Control and Prevention (CDC) is clear: "Individuals should store at least one gallon of water per person per day for at least three days." This recommendation forms the bedrock of our storage strategy, ensuring that in times of crisis, water scarcity is one less worry on your mind.

STORAGE GUIDELINES

Water is fundamental to survival. In emergencies, regular water supply lines can get disrupted, leaving you in dire need. A robust water storage plan is not just a good-to-have; it's a must-have. This section dives into how much water you need to store, explores various storage solutions, and offers insights into selecting the perfect spot for your reserves.

How Much Is Enough?

Following the CDC's advice, storing a minimum of one gallon of water per person per day caters to drinking, cooking, and basic hygiene. However, in hotter climates or for families with special needs, this amount should be adjusted upwards. Planning for a minimum of three days is prudent, but extending that to a two-week supply provides an even greater safety margin.

Container Choices

The vessel you choose for water storage plays a crucial role in ensuring its usability over time. Options range from commercially available water barrels to innovative DIY solutions. Here are a few to consider:

- **Water Barrels:** These are ideal for long-term storage, designed to keep large quantities of water safe. Ensure they're made from food-grade plastic and placed on a sturdy base to prevent any contamination.
- **Portable Containers:** For those with limited space or the need for mobility, containers like jerry cans or water bricks can be invaluable. They're easy to move, stack, and store, even in tight spaces.
- **DIY Solutions:** Repurposing clean soda bottles or other food-safe containers can be an effective and budget-friendly approach. Just ensure they are thoroughly cleaned and disinfected before use.

Choosing the Right Spot

Selecting an optimal location for your water storage is as crucial as the containers you choose. Consider these factors:

- **Temperature Stability:** A cool, dark place is ideal for prolonging the life of your stored water. Extreme temperatures, either hot or cold, can affect water quality over time.
- **Accessibility:** Ensure that your water is easily accessible in an emergency. Avoid storing all your water in hard-to-reach places like a high attic or a cluttered basement.

- **Safety:** Keep your water away from harmful chemicals or areas prone to contamination. Elevated platforms or shelves can prevent contamination from floods or spills.

By adhering to these guidelines, you can establish a foundation of water security that not only meets basic needs but also fortifies your resilience in the face of emergencies. With a well-thought-out plan and the right tools at your disposal, you can ensure that when disaster strikes, your water supply remains uninterrupted, safeguarding the health and well-being of you and your loved ones.

CONTAINER AND QUALITY MAINTENANCE

Ensuring the purity of your stored water is as vital as the air you breathe. Selecting the right containers and maintaining the quality of your water supply are not just steps in the process; they are your lifelines in maintaining a safe, drinkable water supply when the unexpected occurs. Here, we'll delve into the essential practices that will safeguard your water's purity from the moment it's stored until the moment it's needed.

Securing Water Purity

Choosing the Right Containers

The journey to water security begins with the right container—a guardian against contamination. Not all containers are created equal. Your choice must be food-grade, designed to hold water without leaching harmful chemicals or odors into it. This means using containers specifically meant for water storage or those that have previously held food or beverages and are made of materials known for their safety, like certain plastics, glass, or stainless steel.

Maintaining Water Quality

Water, though stored, is not static. Over time, it can become a breeding ground for bacteria if not properly cared for. Here's how to keep your water safe:

- **Regular Rotation:** Refresh your stored water every six months to a year. This practice ensures freshness and minimizes the risk of contamination. Mark the storage date on each container and keep a rotation schedule.
- **Using Preservatives:** For longer-term storage, consider water preservatives specifically designed for emergency water storage. These products can extend the shelf life of your water, ensuring its safety for years.
- **Keeping It Cool and Dark:** Store your water in a cool, dark place to slow down the growth of algae and bacteria. Sunlight and warmth accelerate these processes, compromising water's quality.

Avoiding Common Pitfalls

Even with the best intentions, mistakes can happen. Here are common storage pitfalls and how to steer clear of them:

- **Direct Sunlight:** Sunlight can degrade plastic containers over time and encourage the growth of algae. Keep your water in a dark, cool spot to protect it.
- **Improper Sealing:** Ensure all containers are tightly sealed to prevent the entry of contaminants. Check seals regularly for signs of wear or damage.
- **Contaminant Proximity:** Store your water away from gasoline, pesticides, or any substances that emit fumes. These can permeate plastic containers, contaminating the water.

By emphasizing the importance of container selection and implementing a rigorous water quality maintenance routine, you can ensure your emergency water supply remains safe and reliable. It's not just about having water; it's about having water that's safe to drink, no matter what.

CUTTING-EDGE PURIFICATION

In an age where technology shapes our daily lives, it also empowers us to overcome nature's most pressing challenges. Enter the realm of cutting-edge water purification technologies—a beacon of hope for preppers and survivalists alike. These advanced methods are not just about transforming murky water into a crystal-clear elixir; they represent the forefront of ensuring safety and reliability in your water supply, no matter where you find yourself.

ADVANCED PURIFICATION TECHNOLOGIES

The quest for pure water has led to the development of sophisticated purification technologies, notably UV purification and reverse osmosis. These aren't mere filters; they are shields against the invisible threats that lurk in untreated water, ensuring every drop is safe to drink.

UV Purification: A Ray of Hope

UV (ultraviolet) purification uses the power of light to perform its magic. By exposing water to UV light, it targets the very DNA of bacteria, viruses, and protozoa, rendering them harmless. This technology shines in its ability to neutralize pathogens without adding chemicals to the water or altering its taste. It's efficient, effective, and environmentally friendly.

However, its brilliance does come with considerations. UV purifiers require a source of power, be it from batteries, solar panels, or hand cranks. Maintenance is minimal but crucial; the UV bulb needs regular checks to ensure it is functioning correctly. For preppers, this technology offers a potent tool in their arsenal, especially when stationary or in basecamps where power is accessible.

Reverse Osmosis: Pressing Purity

Reverse osmosis (RO) takes a more brute-force approach. It forces water through a semipermeable membrane, leaving contaminants behind. This process is thorough, removing not just pathogens but also chemicals and heavy metals, delivering water of unparalleled purity.

The trade-offs? Reverse osmosis systems are more complex and often require a pressurized water source. They also necessitate regular maintenance to prevent membrane fouling and to replace filters. Yet, for those willing to navigate these needs, RO provides a level of purification hard to match.

In Action: Real-World Reliance

Case studies abound of these technologies providing lifesaving solutions in disaster-stricken areas, off-grid homes, and international travel. From communities recovering from natural disasters to hikers ensuring their hydration in the wild, these advanced purification methods stand as testaments to human ingenuity in the face of adversity.

DIY PURIFICATION SOLUTIONS

When modern conveniences are stripped away by disaster, or you find yourself far from the nearest clean water source, ingenuity and knowledge become your best allies. DIY water purification solutions embody the spirit of self-reliance, offering lifelines in the form of accessible, effective methods to ensure your water is safe to drink. Here, we'll dive into the how-tos of constructing purification systems, empowering you to harness the power of simplicity to secure one of life's most essential needs.

Solar Water Disinfection (SODIS): Harnessing the Sun

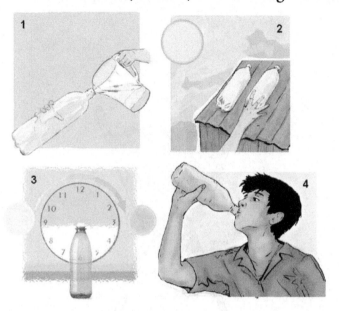

SODIS is a testament to the adage that the best solutions are often the simplest. This method uses sunlight to kill pathogens in water, requiring only clear plastic bottles and sunlight. Here's how to do it:

1. **Choose the Right Containers:** Clear, plastic bottles are ideal. Ensure they are clean and free from scratches.
2. **Fill and Cap:** Fill the bottles with water. If the water is cloudy, filter it through cloth to remove particulates.
3. **Expose to Sunlight:** Place the bottles on a reflective surface under direct sunlight. Six hours of strong sunlight or two days under cloudy conditions are typically sufficient.
4. **Wait and Drink:** After exposure, the water is safe to drink. The UV-A rays and temperature increase in the bottle work together to kill harmful microorganisms.

Homemade Water Filters: A Layered Approach

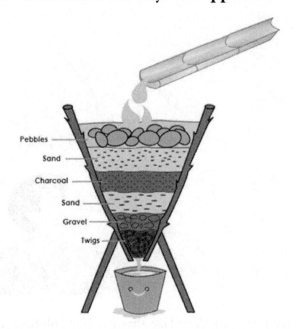

Creating a water filter using sand and charcoal is another effective, low-tech method for purification. This system mimics natural filtration processes, removing sediments and some contaminants:

1. **Gather Materials:** You'll need a container (like a bottle or bucket), charcoal (from a fire, not briquettes), sand, gravel, and a cloth or coffee filter.
2. **Prepare the Container:** Cut the bottom off a bottle or use a bucket with a hole at the bottom. Place the cloth or coffee filter at the bottom to prevent the materials from escaping.
3. **Layer the Materials:** Start with a layer of charcoal, followed by sand, and then gravel. Repeat the layers until the container is filled, ensuring the top layer is gravel.
4. **Filter Your Water:** Pour water through the top and collect it from the bottom. This filtered water should then be boiled or treated with another purification method to ensure safety.

The Importance of Versatility

In emergency scenarios, resources may be limited, and conditions unpredictable. Knowing how to construct DIY purification systems, like SODIS and sand-charcoal filters, equips you with versatile skills that can adapt to various situations. Each method has its strengths and can be deployed based on the materials at hand and the specific challenges faced.

By mastering these DIY purification solutions, you'll not only enhance your survival toolkit but also cultivate a deeper connection with the resourcefulness that has ensured human survival through the ages. This knowledge, paired with practice, can make the difference between vulnerability and resilience, between uncertainty and preparedness.

SELF-RELIANCE TECHNIQUES

Embracing self-reliance means looking to the sky—not in hope, but in strategy. Rainwater collection systems stand as a testament to the ingenuity of those prepared to secure their water independence. These systems don't just capture water; they harness the cycles of nature to provide a renewable source of life-giving fluid, demonstrating a harmony between human needs and environmental stewardship.

RAINWATER COLLECTION SYSTEMS: A DROP FROM ABOVE

The benefits of setting up a rainwater collection system are as clear as the water it gathers. Not only does it offer an additional water source, but it also promotes sustainability and reduces reliance on municipal systems. However, before diving into installation, it's crucial to navigate the considerations that ensure its success and legality.

Designing Your System

Creating an effective rainwater collection system starts with understanding your catchment area—typically your roof. Calculating this area provides insight into the potential volume of water you can collect. For every square foot of catchment area, 0.6 gallons of rainwater can be collected for every inch of rainfall. This calculation is the cornerstone of designing a system that meets your needs.

Storage Solutions

Once collected, where will this water be stored? Options range from barrels for small-scale gardening to large cisterns for household use. Whatever the choice, ensure it is made of materials safe for storing water and placed in a location that protects the water from contamination and excessive temperature fluctuations.

Filtration and Safety

Before use, rainwater should be filtered and purified. Basic filtration can remove debris and sediments, while further purification—such as with UV light or chemical treatment—makes it safe for drinking.

Legal Landscape

Rainwater collection is subject to local laws and regulations, which vary widely. Some areas encourage it with incentives; others have restrictions to manage water rights and environmental impact. Before setting up your system, understand the legal framework in your area to ensure compliance and avoid penalties.

Best Practices

- **Maintain Cleanliness:** Regularly clean your catchment surface and storage containers to prevent contamination.

- **Use First Flush Diverters:** These devices discard the initial rainwater, which may carry pollutants from the catchment surface, ensuring only cleaner water is stored.
- **Monitor Water Quality:** Regularly check your stored rainwater for quality, especially if it will be used for drinking or cooking.

Rainwater collection systems exemplify the essence of self-reliance, merging the practicality of securing a vital resource with the responsibility of sustainable living. By incorporating these systems into your preparedness plan, you not only guarantee an additional water supply but also contribute to a larger cycle of conservation and resilience.

EXTRACTING WATER FROM NATURAL SOURCES

Venturing into the wilderness, whether by choice or circumstance, calls for a deep understanding of how to quench your thirst from the world around you. Nature provides, but it's our wisdom and skills that unlock the resources necessary for survival. Identifying and extracting water from natural sources is not just about finding water—it's about ensuring its safety and sustainability.

Unveiling Nature's Hydration

The first step in sourcing water is knowing where to look. Streams, rivers, and lakes are obvious choices, but what about when those aren't visible? Groundwater might seep out where rock formations meet valleys or at the foot of cliffs. Areas with lush vegetation are likely to have water nearby, as plants need it just as much as humans do.

Capturing Dew and Rainwater

Dew forms on surfaces overnight and can be collected early in the morning using a cloth or sponge, then squeezed into a container. Rainwater, when it comes, can be caught directly or funneled from large leaves or tarps into storage vessels. Both sources offer relatively clean water, but purification is still a prudent step.

Purification Is Key

Water from natural sources carries unknowns. Microorganisms, parasites, and pollutants can pose significant health risks. Boiling is the most straightforward purification method—bring water to a rolling boil for at least one minute (or three minutes at higher altitudes). For situations where boiling isn't possible, chemical treatments—with iodine or chlorine—or UV light purifiers offer alternative solutions. Always let treated water sit for the recommended time before consumption.

Testing Before Trusting

Whenever possible, test the water. Simple test kits can reveal the presence of harmful bacteria or chemicals, providing an extra layer of security. Remember, clear water isn't always clean water.

Extracting water from the environment demands respect for nature and a commitment to leaving no trace. By using these techniques, you can ensure not only your survival but also the preservation of the natural sources we so heavily depend on. This knowledge can empower you to step confidently into the wild, assured in your ability to find and purify the most vital of resources: water.

Now is the moment to look at your water security plan with a critical eye. Have you accounted for storage, purification, and self-reliance? The empowerment that stems from knowing you can provide clean, safe water for yourself and your loved ones in any scenario is unmatched. Take the steps today to elevate your preparedness, and sleep soundly tonight knowing that, come what may, your water needs are covered.

With water security in hand, are you ready to expand your preparedness horizons? The next chapter ventures into the essentials of food security and shelter building—equally critical elements of survival. How would you ensure your food supply if grocery shelves were bare? Can you create a safe haven in the midst of chaos? Join us as we explore these questions, equipping you with the knowledge to thrive in any situation.

Access to a safe water supply is essential for survival. Discover more strategies inside your "DIY Projects for Self-Reliance" bonus resource. Scan the QR code with your smartphone to access this and 9 more bonuses. Alternatively, you can visit this website: https://rebrand.ly/scan-for-your-bonuses

PART 3:
HEALTH AND SAFETY

CHAPTER 7
COMPREHENSIVE SURVIVAL MEDICINE

"In the midst of chaos, there is also opportunity."

— *Sun Tzu*

The wisdom of ancient strategists reminds us that preparedness turns crisis into opportunity. This truth rings especially loud in the realm of survival medicine, where knowledge and readiness can mean the difference between life and despair. This chapter unfolds the tapestry of medical preparedness, weaving through the essentials of assembling the ultimate first-aid kit, mastering lifesaving emergency medical skills, and embracing the power of natural remedies.

Health and wellness become paramount when civilization's comforts are peeled away. Here, you're invited to explore the cornerstones of survival medicine, beginning with constructing a first-aid kit that surpasses basic needs and addressing injuries and illnesses that might arise when professional medical help is not a whisper away. This kit will be your first line of defense, a beacon of hope and safety amidst uncertainty.

But tools alone won't suffice. The heart of survival medicine beats in the skills and knowledge that will empower you to use these tools effectively. From stitching wounds to setting bones, this chapter doesn't just list actions but teaches the how and why behind each one. It's about transforming you from a bystander to a caretaker, equipped to face medical emergencies head-on.

Venturing beyond the conventional, we'll explore the rich world of natural remedies. The earth offers a bounty of healing resources, from the anti-inflammatory properties of willow bark to the antiseptic power of honey. These natural allies can extend your medical toolkit, offering solutions when modern medications are out of reach.

ULTIMATE FIRST-AID KIT

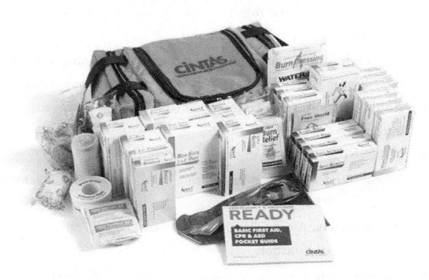

A well-stocked first-aid kit is not just an accessory; it's an essential component of survival preparedness. In the wilderness or the aftermath of a disaster, this kit is often your first and best line of defense, offering tools to manage health emergencies when professional medical care may be hours or days away. What follows is a comprehensive checklist designed to prepare you for a wide range of situations, ensuring you can respond with confidence to injuries, illnesses, and other medical needs that arise in challenging environments.

COMPREHENSIVE CHECKLIST

At the core of your survival strategy should be a first-aid kit tailored not just to basic needs but equipped for long-term survival and diverse scenarios. This checklist, inspired by recommendations from the Mayo Clinic and expanded for comprehensive preparedness, outlines the essentials for a robust medical kit:

- **Bandages and Dressings:** Include a variety of sizes of adhesive bandages, sterile gauze pads, and rolls for dressing wounds. Also, have on hand elastic bandages for sprains and strains.
- **Antiseptic Wipes and Solutions:** Essential for cleaning wounds and preventing infection. Hydrogen peroxide and alcohol pads are staples.
- **Antibiotic Ointment:** To apply on minor cuts and scrapes to ward off infections.
- **Medical Tape:** For securing bandages and gauze in place.
- **Scissors and Tweezers:** For cutting tape, bandages, and removing splinters or ticks.
- **Pain Relievers:** Aspirin, ibuprofen, or acetaminophen can manage pain and reduce inflammation.
- **Antihistamines:** For allergic reactions, bee stings, and insect bites.
- **Thermal Blanket:** To retain body heat in cases of shock or when in cold environments.
- **Gloves:** Disposable gloves to prevent contamination and protect both the caregiver and the patient.
- **CPR Mask:** To safely perform resuscitation.
- **Tourniquet:** For severe bleeding that can't be controlled by pressure.
- **Splinting Materials:** For immobilizing fractures or sprains.
- **Instructions for Emergency Situations:** A compact guide that provides basic first-aid instructions.

Expanding beyond the basics, consider adding items tailored to specific needs or potential scenarios you might encounter, such as snake bite kits, water purification tablets, or personal medications.

BEYOND CONVENTIONAL MEDICINE

In the quest for comprehensive preparedness, venturing beyond the realm of conventional medicine opens up a world of possibilities. This exploration is not about replacing traditional medical supplies but enhancing your first-aid arsenal with unconventional resources. Veterinary antibiotics and natural remedies stand out as potent allies, offering alternative solutions when standard medical help might not be accessible.

Veterinary Antibiotics: A Contingency Plan

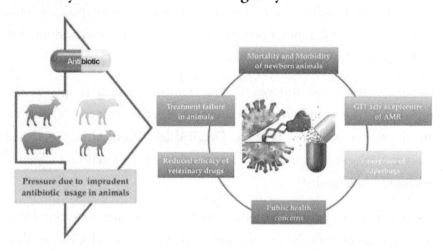

The idea of including veterinary antibiotics in a human survival kit might raise eyebrows, yet in extreme situations, they can be invaluable. These medications often share active ingredients with their human counterparts, offering a bridge to treatment when pharmacies are out of reach. The key to their inclusion is understanding their composition and the specific conditions they can treat. Safety is paramount. It's crucial to research and document the exact dosages, potential side effects, and contraindications of any veterinary medication considered for human use. Organizations like the World Health Organization have published resources on the interchangeability of certain veterinary and human antibiotics, providing a foundation for making informed decisions.

Natural Remedies: Earth's Healing Touch

Natural remedies, derived from plants and essential oils, offer a complementary approach to health care, especially in survival scenarios.

Their efficacy in treating various ailments, from wounds and infections to pain and inflammation, has been recognized in numerous studies.

For instance, honey, with its antibacterial properties, can be applied to cuts and burns to prevent infection and promote healing. Similarly, lavender oil, known for its calming effects, can also serve as an antiseptic for minor wounds. Tea tree oil, another powerful natural antiseptic, can be diluted and used for skin infections or insect bites.

The preparation and storage of these remedies are as simple as their ingredients. Many can be stored in compact, sealed containers and kept cool and dry to preserve their potency. Detailed labels indicating the remedy, concentration, and suggested uses can ensure they are safely and effectively employed when needed.

Incorporating veterinary antibiotics and natural remedies into your survival kit is a testament to the innovation and adaptability crucial for preparedness. By embracing these unconventional resources, you'll significantly broaden your medical response capabilities, ensuring you're equipped to face a wider range of health challenges with confidence.

EMERGENCY MEDICAL SKILLS

In the silence that follows a crisis, when professional medical help is beyond reach, the knowledge and application of advanced first-aid techniques can shine as a beacon of hope. These skills bridge the gap between immediate needs and professional care, often becoming the critical difference in survival situations. This section delves into the art and science of emergency medical interventions, offering a lifeline through step-by-step tutorials on essential procedures.

ADVANCED FIRST-AID TECHNIQUES

Suturing Wounds

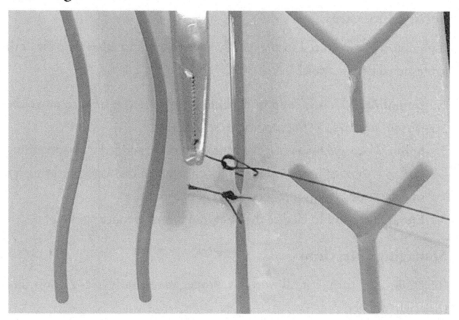

A deep cut in an environment far from a hospital requires prompt action to prevent infection and promote healing. Here's how to suture a wound in a step-by-step manner:

1. **Clean:** First, wash your hands and wear gloves. Clean the wound thoroughly with saline solution or clean water.
2. **Anesthetize:** If possible, apply a local anesthetic around the wound to reduce pain.
3. **Align:** Carefully bring the edges of the wound together. Use forceps if necessary.
4. **Suture:** Using a sterilized needle and thread, insert the needle about a quarter-inch from the edge of the wound, then pull it across to the other side, exiting at an equal distance from the edge. Knot the suture to hold the wound closed. Repeat along the wound, spacing stitches evenly.

5. **Dress:** Once suturing is complete, cover the wound with a sterile dressing.

Setting Fractures

Immobilizing a fracture can prevent further injury and alleviate pain until professional help is available.

1. **Immobilize:** Gently support the injured area. If you have to move the person, stabilize the fracture with a splint.
2. **Splint:** Use rigid materials, like sticks or boards, padded with something soft, to create a splint. Bind it with cloth strips or bandages, ensuring it's snug but not so tight that it cuts off circulation.
3. **Elevate:** If possible, elevate the fractured limb to reduce swelling.

Managing Infections

Infections can turn critical without proper care. Early recognition and treatment are key.

1. **Identify:** Be on the lookout for redness, swelling, heat, and pain—signs of infection.
2. **Clean:** Regularly clean the area with a gentle antiseptic solution.
3. **Antibiotics:** If available and appropriate, administer antibiotics following specific guidelines for the type of infection.

Accurate, reliable information underpins all these techniques. Resources like the American Red Cross and medical textbooks provide a solid foundation for learning and applying these skills. Remember, practicing under the guidance of professionals and within legal boundaries is essential to ensure that when the moment comes, you're prepared not just to act but to act wisely and effectively.

SELF-RELIANCE IN MEDICAL EMERGENCIES

Empowerment in medical emergencies springs from the confidence and competence to act decisively when seconds count. This self-reliance is not born from isolation but from the strength of knowledge and the ability to apply critical medical interventions on one's own. Here, we'll explore the essential, hands-on techniques that everyone should master: CPR, the Heimlich maneuver, and the principles of basic triage. These skills are lifelines, turning ordinary individuals into heroes in moments of crisis.

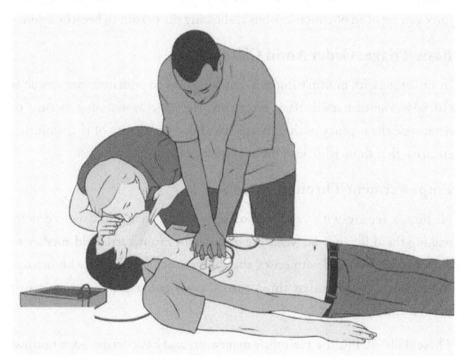

CPR: The Beat of Life

Cardiopulmonary resuscitation (CPR) is a fundamental skill that can double or even triple a cardiac arrest victim's chance of survival. The sequence is straightforward:

1. Check responsiveness and breathing.
2. Call for emergency help if the person is not breathing.

3. Place your hands one on top of the other, in the center of the chest. Push hard and fast, at least 2 inches deep, and at a rate of 100 to 120 compressions per minute.

4. If trained, provide rescue breaths after every 30 compressions.

The Heimlich Maneuver: Clearing the Way

Choking can be deadly, but the Heimlich maneuver offers a powerful tool to save lives. The process involves standing behind the person, wrapping your arms around their waist, and giving quick, upward abdominal thrusts. This force can expel an obstructing object, allowing the person to breathe again.

Basic Triage: Order Amid Chaos

In situations with multiple injuries, knowing how to prioritize care can be as crucial as the care itself. Basic triage involves quickly assessing victims to determine the urgency of their needs based on the severity of their injuries, ensuring that those who need immediate attention receive it first.

Empowerment Through Action

History is replete with examples of self-reliance in medical emergencies making the difference between life and death. From battlefield medics to everyday citizens faced with crises, the ability to act, informed by knowledge and practice, has countless times turned potential tragedy into stories of survival and hope.

These skills—CPR, the Heimlich maneuver, and basic triage—are not just techniques but are expressions of the human capacity to care for one another in the most direct way possible. Learning and practicing them will equip you with more than the means to save a life; they will empower you with the confidence that comes from knowing you can stand firm in the face of medical emergencies, ready to offer hope and help when it's needed most.

NATURAL REMEDIES

In the vast pharmacy of nature, plants and oils offer a treasure trove of healing powers waiting to be unlocked. The wisdom of herbal medicine and the potency of essential oils complement conventional medical practices, particularly in situations where traditional resources may be scarce. This section dives into the heart of natural remedies, guiding you through the medicinal properties of flora and fauna that have stood the test of time, backed by scientific research and generations of traditional use.

MEDICINAL USES OF PLANTS AND OILS

Nature's bounty provides not just food but medicine, offering solutions for everything from soothing cuts and burns to battling infections and reducing inflammation. Here, we'll explore the potent allies that can be included in your survival medicine kit.

- **Aloe Vera:** Known for its soothing and healing properties, aloe vera gel can be applied to burns, cuts, and other skin injuries to promote healing and reduce inflammation.
- **Lavender Oil:** Beyond its calming scent, lavender oil has antiseptic and anti-inflammatory properties. It can be applied to minor burns and bug bites or used in a diffuser to relieve stress.
- **Chamomile:** As a tea or applied topically, chamomile has been shown to soothe skin irritations, reduce anxiety, and aid sleep. Its gentle action makes it suitable for all ages.
- **Tea Tree Oil:** This powerful antimicrobial can be diluted and applied to cuts, scrapes, and fungal infections. It's a must-have for its broad-spectrum antibacterial and antifungal benefits.
- **Peppermint:** Whether used as an oil or fresh leaves, peppermint can relieve headaches when rubbed on the temples, improve digestion, and refresh the senses.

Cultivating a basic understanding of these natural resources, along with knowledge on their proper collection, preparation, and storage, will enrich your survival toolkit. Learning to recognize these plants and how to harvest their active components will ensure that you can tap into their healing potential.

If you're looking to integrate these natural remedies into your preparedness plan, consider starting a small herb garden or keeping a few essential oils on hand. This will not only ensure access to natural medicines but will also foster a deeper connection with the healing powers of the natural world. Whether you're dealing with minor ailments or supporting overall well-being, the inclusion of natural remedies in your survival strategy will offer you versatility and resilience in facing the challenges ahead.

PRACTICAL APPLICATIONS

Harnessing the healing power of nature not only requires knowledge of medicinal plants and oils but also an understanding of how to transform them into practical remedies. The alchemy of preparing and storing these natural gifts ensures their potency and safety, readying them for the moment they're needed. This segment offers a blueprint for crafting herbal remedies and essential oil blends, complete with guidelines for their preservation and safe use.

Crafting Herbal Remedies and Essential Oil Blends

Creating effective natural remedies is an art grounded in science. Here are a few simple recipes to begin with:

- **Calendula Salve for Skin Healing:** Melt 1 part beeswax in 5 parts calendula-infused oil over a double boiler. Once fully melted, pour into containers and let cool. Harnessing calendula's natural anti-inflammatory and healing properties, this salve is excellent for cuts, scrapes, and chapped skin.

- **Peppermint and Lavender Headache Balm:** Combine equal parts of coconut oil and shea butter with a few drops of peppermint and lavender essential oils. Store in a small jar and apply to temples and the back of the neck for headache relief.

- **Echinacea Tincture for Immune Support:** Fill a jar with dried echinacea and cover it with vodka, ensuring the plant material is completely submerged. Seal the jar and store it in a cool, dark place for 4 to 6 weeks, shaking daily. Strain the mixture through cheesecloth and store the liquid in a dark dropper bottle. A few drops can be taken daily during cold and flu season to boost the immune system.

Preserving and Storing Natural Medicines

The efficacy of natural remedies depends significantly on their proper storage. Here are key strategies to ensure longevity:

- **Dark, Cool Storage:** Light and heat degrade active compounds. Store dried herbs and essential oil blends in dark glass containers in a cool, dry place.
- **Airtight Containers:** Oxygen can spoil natural remedies. Use airtight containers to minimize exposure and preserve potency.
- **Labelling:** Always label your remedies with their name, ingredients, preparation date, and dosage instructions. This ensures safe and effective use.

By mastering the preparation and storage of natural remedies, you can empower yourself with self-sufficiency in health care. This knowledge, paired with a respect for nature's potency and an adherence to safety practices, will ensure that your foray into the world of natural medicine is both beneficial and secure.

Now is the time to transform insight into action. Equipped with the knowledge of survival medicine, from the essentials of a first-aid kit to the wisdom of natural remedies, the power to safeguard your health in any situation rests in your hands. Let this chapter be your springboard into a state of readiness, where preparedness meets practice.

As you turn the page, the journey deepens into the realms of psychological resilience and advanced survival strategies. The next chapter unveils the secrets to not just surviving but thriving in the face of adversity. Mastering these strategies is paramount to building a comprehensive readiness plan, ensuring you're equipped to face whatever the future holds with confidence and strength.

Thank you for purchasing, "The Complete Prepper's Survival Bible." Scan the QR code to claim your 10 bonuses. You'll find everything from essential medical information to handling emergencies off-grid. Alternatively, you can visit this website:

https://rebrand.ly/scan-for-your-bonuses

CHAPTER 8
SHELTER AND SAFETY STRATEGIES

"A shelter is more than a roof over your head—it's your frontline
defense in survival."

— *John Wiseman, SAS Survival Handbook.*

Safety and shelter are not just basic needs; they are the foundation of any effective survival strategy. Whether facing natural disasters, societal upheaval, or just the uncertainties of life, having a secure place to call home offers more than comfort—it offers a tactical advantage. This chapter dives deep into the essentials of constructing a sturdy, reliable shelter that stands up to various challenges, fortifying it against intruders, and, if push comes to shove, how to defend it with more than just strong words.

We'll walk through the steps of designing shelters that blend into their surroundings, offering safety without sacrificing sustainability. From selecting the right materials to understanding the terrain and climate you're working with; we'll lay down the blueprints for survival homes that do more than stand up—they stand out for their ingenuity and resilience.

Next, we'll shift our focus to perimeter defense. It's not just about keeping out the bad—it's about creating a buffer zone that gives you the upper hand. Through smart design and strategic planning, you'll learn how to use the environment to your advantage, making any unwelcome guests think twice before stepping foot on your turf.

But what if the worst happens? We don't stop at prevention. This chapter also explores improvised weaponry—tools you can create from everyday materials when there are no other options. It's a guide to being resourceful, prepared, and aware of the legal and ethical considerations that come with self-defense.

By the end of this chapter, you'll be equipped not just with knowledge but also with the confidence to build, defend, and sustain a shelter that keeps you safe, rain or shine. Let's get started on making safety and security your top priorities, creating peace of mind in an uncertain world.

ADVANCED SHELTER DESIGN

In the quest for safety, the art of building a shelter transcends mere survival; it becomes a statement of resilience and foresight. A well-crafted shelter not only provides a haven from the harshness of the elements but also a bastion of security in uncertain times. This section delves into the nuances of constructing shelters tailored to withstand the tests of time and nature, focusing on the blend of durability and adaptability.

BUILDING INSTRUCTIONS

The cornerstone of any enduring shelter lies in its design and the materials chosen. The process begins with understanding the environment you're in. Are you facing the biting cold of the north, the searing heat of the desert, or the unpredictable conditions of a temperate climate? Each scenario demands a unique approach, a different palette of materials, and a distinct architectural strategy.

Step 1: Site Selection First, choose a location that leverages natural protection. Look for natural windbreaks in cold environments, elevation in flood-prone areas, and shade in hot climates. Your site is your first layer of defense.

Step 2: Design According to Climate In colder regions, consider a partially underground design, utilizing the earth's natural insulation. For warmer climates, elevated structures with wide overhangs promote air circulation, reducing heat inside. Each design should consider the sun's path, prevailing winds, and local wildlife.

Step 3: Material Selection Opt for locally sourced, durable materials. In forested areas, wood might be abundant, but remember, it requires treatment to withstand the elements. In contrast, areas devoid of trees might necessitate the use of mud bricks or stone. Incorporating sustainable resources, such as rainwater collection systems or solar panels, enhances your shelter's long-term viability.

Step 4: Incorporate Resilience Features Think about fire-resistant materials in fire-prone areas or sturdy, wind-resistant designs in hurricane or tornado zones. Features like a reinforced shelter roof can double as a water collection system, turning a basic need for protection into an advantage.

Expert Insights John Smith, in his 2023 study on sustainable shelter constructions, highlighted the importance of choosing materials not just for their immediate availability or cost but for their longevity and environmental impact. He noted, "Sustainable building materials are not a luxury but a necessity in constructing shelters that last. Materials like bamboo, adobe, and recycled steel offer durability, efficiency, and are less taxing on the environment."

Through careful planning, an understanding of the environment, and strategic material selection, building a shelter that stands the test of time is not only possible but essential for long-term survival. This approach will ensure your shelter is not just a place to live but a stronghold of safety and sustainability.

SHELTER BLUEPRINTS: CRAFTING YOUR SURVIVAL MASTERPIECE

In the world of survival, blueprints are not just drawings—they are your roadmap to safety, efficiency, and resilience. A well-thought-out blueprint does more than lay the foundation for a sturdy shelter; it integrates your survival strategy into the very walls that protect you. Let's delve into the art and science of creating blueprints for two of the most strategic shelters: underground bunkers and treehouses. These examples illustrate not just the shelter itself but how foresight in planning can turn a good shelter into an unbeatable fortress.

Underground Bunkers: Subterranean Safety

Imagine a shelter so discreet that it's invisible to the untrained eye, yet so robust it can withstand extreme conditions. Underground bunkers offer unparalleled protection from the elements and unwanted visitors. But there's an art to their design. It begins with selecting a location: high enough to avoid water table issues, yet accessible enough for construction and emergency exit.

The blueprint for an underground bunker must account for ventilation, water filtration, and sustainable power sources. Solar panels discreetly placed away from the main entrance and camouflaged can power underground life support systems. Rainwater harvesting systems integrated into the design can provide a sustainable water supply.

Consideration of the living space is crucial. Compact design principles can make a small area feel spacious and functional. Incorporating storage into the walls, a multipurpose area for sleeping and communal activities, and an efficient kitchenette can make underground living not just bearable but comfortable.

Treehouses: High and Mighty

Sometimes, elevation is your best defense. Treehouses, often seen as childhood escapes, can be formidable shelters with the right design. The blueprint for a survival treehouse focuses on camouflage, mobility, and defense.

Choosing the right tree is the first step—strong, with a wide canopy for cover, yet with clear sightlines for observation. The design should include a simple pulley system for supplies, retractable ladders for access control, and insulated walls for weather protection and to blend in with the natural surroundings.

Treehouses offer strategic advantages, such as early warning of approaching threats and natural barriers against intruders. Designing for camouflage, using materials that blend with the environment, and incorporating natural defense features, like surrounding thorny plants, enhance these advantages.

Crafting Your Blueprint

Creating a blueprint requires thoughtful consideration of your environment, needs, and potential threats. Architectural design for survival shelters goes beyond the traditional, integrating strategic elements that cater to both immediate and long-term survival.

Experts in survival architecture, like Smith (2023), advise on the importance of energy efficiency and resource management in shelter design. Modifications such as passive solar heating, thermal mass insulation, and rainwater catchment systems are not just environmentally friendly; they are essentials for long-term sustainability in a survival situation.

Whether you're drawing up plans for a subterranean hideout or a treetop haven, remember that your blueprint is more than a plan. It's a declaration of resilience, a testament to your commitment to survival, and your first step toward turning the tide in your favor. With careful planning, a bit of ingenuity, and a keen eye for detail, you can create a shelter that's not just a place to stay but a place to thrive.

PERIMETER DEFENSE: THE INVISIBLE SHIELD

Securing the perimeter of your sanctuary is not just about erecting walls; it's about weaving a web of safety that blends seamlessly with the environment, offering both deterrent and defense without broadcasting its presence. The true art of perimeter defense lies in its subtlety and in the strategic integration of natural and technological elements. Here, we'll explore how to fortify your haven, ensuring it remains your stronghold against any intrusion.

SECURITY MEASURES: CRAFTING A COMPREHENSIVE SHIELD

The first line of defense is often the most overlooked nature itself. Natural barriers can be your silent guardians. Thorny plants and dense shrubs not only serve as a prickly welcome for unwelcome guests but also add to the aesthetic and ecological value of your surroundings. Water features, such as moats or natural streams, can be incorporated into your landscape design, offering both beauty and a formidable obstacle to those on foot.

Visibility plays a crucial role in security. Clear lines of sight around your perimeter will allow you to spot potential threats from a distance, while strategic landscaping can hide your shelter from prying eyes. The goal is to see without being seen, maintaining an advantage over any who might approach with ill intent.

Accessibility is a double-edged sword. While your shelter needs to be reachable in times of need, it should not welcome unwanted access. Simple solutions, like hidden paths, disguised entrances, or even decoy trails, can mislead and deter. Design your access points with care, ensuring they are secure yet accessible to you and your trusted circle.

Defensibility extends beyond physical barriers. Advanced surveillance systems—once the domain of high-security facilities—are now accessible and adaptable for personal use. Solar-powered cameras, motion sensors, and remote monitoring systems can be discreetly placed around your perimeter, offering high-tech oversight without the need for conspicuous power sources.

Integrating these measures into your shelter's design requires foresight and creativity. For instance, consider how the natural growth of thorny plants can be guided to create a living fence, or how surveillance cameras can be camouflaged within birdhouses or tree trunks. The key is to blend your defenses into the environment so well that they go unnoticed until they're needed.

Remember, the strength of your perimeter defense doesn't solely lie in its components but in how seamlessly and intelligently they are integrated into your overall survival strategy. It's about creating a balance that provides protection without feeling like a fortress, ensuring your safety while preserving the peace and beauty of your sanctuary. By adopting a layered approach, combining natural deterrents with technological surveillance, you can create a perimeter that is as elegant as it is effective, offering peace of mind in an uncertain world.

CONTINGENCY PLANNING: THE ART OF PREPARED ESCAPE

When it comes to survival, hope for the best but plan for the worst. The foundation of any robust survival strategy isn't just how well you fortify your haven but how prepared you are to leave it when necessary. Contingency planning is about foreseeing the unforeseeable, creating a blueprint for safety that extends beyond the boundaries of your shelter. This section delves into the critical aspects of escape routes and evacuation plans, emphasizing their role as lifelines in times of crisis.

Mapping Your Escape: Routes and Routines

The core of effective contingency planning lies in the meticulous mapping of escape routes. These should be designed with the understanding that the nature of threats is as varied as the environment itself. Fires, floods, hostile encounters, or even structural failures—each scenario demands a unique response plan. Begin by identifying multiple exit points from your shelter, ensuring that each route is viable under different conditions.

Escape routes should not only lead away from immediate danger but also toward predetermined safe zones. These zones could be anything from a secondary shelter to a rendezvous point with other members of your community. The key is clarity and accessibility, ensuring that, even under stress, the path to safety is unmistakable.

Rehearsing the Unthinkable

Knowing the way out is only half the battle; the other half is practice. Regular drills are essential, ensuring that every inhabitant of the shelter can execute evacuation plans with their eyes closed. These rehearsals should mimic realistic conditions as closely as possible, incorporating potential obstacles and variables. It's not just about speed but also about making smart choices under pressure.

Flexibility: The Unsung Hero of Planning

A plan that cannot adapt is a plan that will fail. Flexibility is the golden rule of contingency planning. As situations evolve, so must your escape strategies. This means regularly reviewing and updating your plans in light of new information, changes in your environment, or alterations to your shelter.

Involve every member of your shelter in the planning process. A diverse perspective can unearth potential oversights and foster a sense of ownership and understanding among all involved. Empower each person with the knowledge and confidence to make critical decisions when the plan needs to deviate.

In the grand scheme of survival, your greatest asset is not the walls that protect you but the wisdom to navigate away from danger. Contingency planning is about embedding this wisdom into every brick, every path, and every decision, ensuring that when the world outside becomes too hostile, you have a way out, not just a way to hide. By embracing flexibility, prioritizing rehearsal, and mapping your escape, you prepare not just to survive but to thrive, no matter what challenges lie ahead.

IMPROVISED WEAPONRY: THE ART OF RESOURCEFUL DEFENSE

In the realm of survival, the ability to defend oneself is paramount. However, not all situations allow for traditional means of protection. Herein lies the value of improvised weaponry—tools fashioned from everyday items, serving as a testament to human ingenuity and resilience. This section is dedicated to the craft of turning the ordinary into the extraordinary, emphasizing that the true weapon is not the object itself, but the knowledge and skill behind its creation and use.

FORGING TOOLS OF SURVIVAL: CREATION AND USE

The essence of improvised weaponry lies in its simplicity and accessibility. Items that are commonplace in any household or environment can become instruments of defense in skilled hands. The key is to view your surroundings through the lens of potential: What items can be combined? Which materials possess inherent strength or can be easily manipulated?

For instance, a broomstick can transform into a spear with the addition of a knife or sharp object affixed to one end. Duct tape, a survivalist's staple, can secure the blade in place. This spear can keep an attacker at a distance, offering a strategic advantage with minimal risk to the wielder.

Similarly, a sling made from a scarf or piece of fabric can launch small rocks or objects with surprising velocity, serving as a ranged weapon when close combat is not favorable. The effectiveness of such improvised weapons relies on the element of surprise and the skill of the user.

However, with the power of these makeshift defenses comes responsibility. The creation and use of improvised weapons are accompanied by inherent risks, not only to potential aggressors but to the user and bystanders. It's crucial to practice caution and restraint, understanding the legal and moral implications of defensive actions.

Training and practice are the foundations upon which the effectiveness of improvised weapons is built. Familiarity with the weight, balance, and force required for each item ensures that, if needed, your response will be both confident and controlled. Regular practice refines the skills necessary to use these tools effectively, turning desperation into a display of preparedness and resolve.

Improvised weaponry is a last resort, a symbol of the survival spirit's adaptability and resourcefulness. In understanding and respecting the power of these makeshift tools, one embodies the essence of survival: the will to persevere, protect, and prevail against all odds.

LEGAL AND ETHICAL ISSUES: NAVIGATING THE FINE LINE

In the crucible of survival, the line between defense and aggression can blur. While the instinct to protect oneself and loved ones is fundamental, it's crucial to tread this path with both legal awareness and ethical consideration. The use of improvised weaponry, while ingenious, introduces a complex layer of responsibility that one must navigate with care. This section explores the delicate balance between survival and societal norms, aiming to equip readers with the knowledge to make informed decisions in the heat of the moment.

Understanding the Legal Landscape

The legalities surrounding self-defense and the use of improvised weapons vary significantly from one jurisdiction to another. What's deemed acceptable in one area could lead to serious legal consequences in another. It is paramount for survivalists to not only be resourceful but also informed. Familiarizing yourself with local laws regarding self-defense is not just advisable; it's essential.

For example, many regions allow for the use of force in self-defense but only to the extent necessary to neutralize an immediate threat. The introduction of a weapon into a conflict, even one improvised, can escalate the legal scrutiny of your actions. Documenting and understanding the legal framework in your area provides a guidepost for both preparation and action.

Ethical Considerations and Responsibility

Beyond the letter of the law lies the realm of ethics. The decision to use force, particularly lethal force, carries weight beyond the immediate consequences. Ethical self-defense involves assessing the situation critically, considering the necessity and proportionality of your response.

In moments of danger, the distinction between defense and excess can become clouded. Training and mental preparedness play key roles in ensuring that if you must act, you do so with a clear understanding of both the immediate and rippling effects of your actions.

Consulting with legal experts or engaging with local law enforcement can provide insights into the practical application of self-defense laws. Workshops or community programs focused on legal and ethical self-defense offer valuable perspectives, preparing you to act not just with confidence but with conscience.

In the landscape of survival, where every decision can have lasting implications, the importance of legal and ethical preparedness cannot be overstated. It is a component of survival strategy that fosters not only personal safety but also the integrity and moral responsibility of the individual.

You've journeyed through the essentials of shelter and safety, uncovering the layers of preparation that fortify your resilience against any storm. The knowledge you've gained is more than just words on a page—it's the blueprint for empowerment, the foundation upon which you can build a fortress of security and peace. Now is the moment to move from contemplation to action, to weave these strategies into the fabric of your daily life. Let the insights on advanced shelter design, perimeter defense, and even the prudent use of improvised weaponry guide you in crafting a haven that stands vigilant against the unforeseen.

As we close this chapter, remember that preparedness is a continuous journey, one that evolves with every new skill acquired and every piece of knowledge gained. But the path doesn't end here. Looking ahead, we venture into realms that will further cement your autonomy and readiness— advanced survival skills await, promising deeper insights into self-sufficiency. How will you build a community that thrives on resilience? What strategies will ensure your living practices are not just sustainable but regenerative?

The next chapter beckons, ready to explore these questions and more, inviting you to elevate your preparedness to new heights. Embrace the journey ahead. Every step taken is a step toward unwavering preparedness, toward a life where confidence overshadows fear, and readiness turns challenges into opportunities. Let's continue forward, together, on this path to comprehensive preparedness.

Shelter is paramount in any crisis. Gain additional prepping insights by claiming your 10 bonuses. Simply scan the QR code above. Alternatively, you can visit this website:

https://rebrand.ly/scan-for-your-bonuses

CHAPTER 9
OFF-GRID LIVING ESSENTIALS

"In the depth of winter, I finally learned that within me there lay an
invincible summer."
— *Albert Camus*

The leap into off-grid living is more than a journey; it's a return to the essence
of what it means to be self-reliant, resilient, and in harmony with the natural
world. It's about forging a life that is not just lived but deeply felt, a life where
every drop of water, ray of sunshine, and breath of wind is a reminder of our
connection to the earth and our ability to adapt and thrive within it.

This chapter lays the foundation for a transition to off-grid living,
emphasizing that the key to success lies not in eschewing modernity but in

embracing a balance between innovation and the timeless wisdom of natural living. It's a path that requires careful planning, preparation, and, above all, a commitment to sustainable practices that nurture both our homes and the environment.

We will explore the critical first steps in this transformative journey: selecting a location that offers the perfect blend of seclusion, access to resources, and sustainability; assembling a bug-out bag that goes beyond mere survival to provide comfort and security in any scenario; and adopting practices that ensure our off-grid life is not just viable but vibrant and fulfilling.

Through strategic planning and a deep respect for the natural world, off-grid living can become not just a dream but a tangible, rewarding reality. It's a testament to our capacity for independence, a reflection of our desire for a life less ordinary, and a commitment to a future where we live in harmony with the world around us. Let's embark on this adventure, equipped with the knowledge, tools, and determination to forge a life that's both sustainable and deeply satisfying.

BUG-OUT LOCATION SELECTION: FINDING YOUR SANCTUARY

Choosing the right bug-out location is the cornerstone of successful off-grid living. This decision will shape your journey, influencing not just your lifestyle but your chances of thriving in a world where being prepared is synonymous with freedom. It's about finding a spot that not only shelters but also nurtures, a place where you can grow roots and stand tall against the challenges of self-sufficiency.

STRATEGIC CRITERIA: MAPPING THE PATH TO INDEPENDENCE

The quest for the perfect bug-out location is guided by key criteria, each serving as a pillar supporting your off-grid dream. These criteria are not just boxes to tick. They are values to uphold, ensuring your chosen haven stands as a bastion of sustainability, security, and serenity.

Access to Natural Resources

Your survival sanctuary must be rich in natural resources. Water is the lifeblood of any off-grid homestead. Proximity to a reliable water source—be it a stream, lake, or a feasible spot for a well—is non-negotiable. Similarly, fertile land for cultivation and forests for wood and forage offer the sustenance and warmth essential for year-round living. Smith (2021) underscores the balance needed between seclusion and resource availability, advocating for locations that do not compromise on either.

Remoteness for Privacy and Security

The allure of off-grid living often lies in its promise of privacy and a life untouched by the chaos of the outside world. A remote location ensures that you're off the radar, away from the prying eyes and the uninvited. However, this remoteness shouldn't translate to isolation. A balance must be struck, ensuring that while your haven is discreet, it remains accessible in emergencies, both to you and for any essential support.

Suitability for Long-term Sustainability

The true test of a bug-out location is its capacity to support long-term sustainability. This means more than just survival; it's about creating a home that can evolve, adapt, and thrive over time. The land should offer not just immediate resources but also the potential for renewable energy sources, such as solar or wind power. The climate should be conducive to year-round living, with seasons that complement your lifestyle rather than challenge it.

Choosing a bug-out location is perhaps the most critical decision in your off-grid journey, laying the groundwork for a life of independence. It requires careful consideration, foresight, and a deep understanding of your own needs and aspirations. By adhering to these strategic criteria, you're not just selecting a place to live; you're choosing a life filled with potential, grounded in the earth, and open to the sky.

CONSIDERATIONS

Venturing into off-grid living demands more than just a love for solitude and sustainability; it requires a keen awareness of the legal tapestry and environmental ethos that govern your chosen sanctuary. The dream of self-sufficiency must be pursued with respect for the laws of the land and the health of the ecosystem. This section delves into the essential legal and environmental considerations needed to ensure your off-grid lifestyle is both legitimate and harmonious with nature.

Legal Framework: Building Within Bounds

The legal landscape surrounding off-grid living is as varied as the terrain itself. Local zoning laws, property rights, and building codes form the framework within which your off-grid dream must fit. Understanding these regulations is paramount to avoid costly missteps and potential legal battles. For instance, some areas may restrict the types of structures you can build or prohibit off-grid living altogether.

Property rights are another critical consideration. Ensuring you have clear title and understanding any easements or restrictions on your land can prevent future disputes. Similarly, being aware of environmental protection regulations helps protect the natural resources vital to your off-grid life. It's advisable to consult with local authorities or legal experts early in the planning process. As Smith (2021) notes, "The ideal off-grid location is one where legal clarity and environmental sustainability intersect."

Environmental Stewardship: Living in Harmony

Choosing to live off-grid is often motivated by a desire to reduce one's ecological footprint. However, good intentions must be matched with responsible environmental stewardship. This includes understanding the impact of your activities on the local ecosystem, from water usage and waste management to land clearing and agriculture.

Sustainable practices, such as rainwater harvesting, solar power, and permaculture, not only align with off-grid ideals but also with the principles of conservation. By designing your off-grid lifestyle around the health of the environment, you contribute to the preservation of the natural world for future generations.

Adapting to the local ecosystem rather than imposing on it ensures a symbiotic relationship between your homestead and the surrounding nature. For example, choosing native plants for your garden supports local wildlife and minimizes water usage, while sustainable building materials reduce your carbon footprint.

The path to off-grid living is as much about navigating the complexities of legal and environmental considerations as it is about the physical construction of a homestead. By approaching these considerations with diligence and respect, you lay the groundwork for a life that is not only self-sufficient but also sustainable and lawful, ensuring your off-grid dream enriches both your life and the land you call home.

COMPREHENSIVE BUG-OUT BAG: YOUR LIFELINE IN A PACK

A well-equipped bug-out bag isn't just a survival kit; it's your lifeline in times of unexpected crisis or when venturing into the wilderness of off-grid living. This carefully curated pack is designed to sustain you for the first crucial 72 hours of an emergency, providing essentials across shelter, water, food, clothing, and tools. Let's unpack the vital components that make up this indispensable survival asset.

ESSENTIAL CONTENTS: THE CORE OF YOUR SURVIVAL KIT

- **Shelter**: Your first line of defense against the elements.

- o Lightweight tent or tarp
- o Emergency space blanket
- o Sleeping bag suitable for the climate
- **Water**: Staying hydrated is non-negotiable.
  - o Water purification tablets or portable filter system
  - o Collapsible water container
  - o Stainless steel water bottle (doubles for boiling water)
- **Food**: Nutrient-rich, non-perishable sustenance to keep you going.
  - o High-energy bars and ready-to-eat meals
  - o Compact fishing kit
  - o Portable stove and fuel, if space permits
- **Clothing**: Adaptability to weather conditions is key.
  - o Layered clothing for temperature regulation
  - o Rain gear and thermal underwear
  - o Durable gloves and hat
- **Emergency Tools**: Equip yourself for any situation.
  - o Multi-tool with knife
  - o Flashlight or headlamp with extra batteries
  - o Firestarter kit (matches, lighter, and waterproof tinder)

Each item in your bug-out bag serves a specific purpose, tailored to support survival in diverse environments. The significance of customizing your bag according to climate cannot be overstated. Whether facing the chill of mountainous terrains or the heat of arid landscapes, your gear must be appropriate for your environment.

Experts and seasoned survivalists underscore the importance of scenario-based adaptations. For instance, if you're in an area prone to natural disasters, like floods or earthquakes, your bag should include additional items, such as a whistle for signaling help or a wrench to turn off gas lines. Studies and guides on emergency preparedness often highlight the necessity of personalizing your bug-out bag to fit not just your needs but also those of your environment. Creating a comprehensive bug-out bag is a dynamic process, requiring regular review and updates as your situation and environment evolve. With each item selected for its utility, durability, and versatility, your bug-out bag becomes more than just equipment—it becomes a testament to your readiness to face the unpredictable, armed with the essentials to survive and thrive.

PERSONALIZATION TIPS: TAILORING YOUR BUG-OUT BAG

A one-size-fits-all approach doesn't apply when it comes to bug-out bags. The true value of this emergency kit lies in its ability to meet the specific needs of you and your family, ensuring that each member can survive and thrive during the first critical hours of a crisis. Personalizing your bug-out bag is about more than just preference; it's about practicality, ensuring that every item within your pack has a purpose tailored to your unique situation.

Assessing Individual Needs: The Foundation of Personalization

- **Medical Requirements**: Health considerations are paramount. Include prescription medications, with a focus on a supply that can last several days. Consider allergies and include appropriate antihistamines or

EpiPens. A personalized first-aid kit should also cater to known medical conditions within your family, such as asthma or diabetes.

- **Dietary Preferences**: Energy is essential in survival situations, making it crucial to pack food that matches your dietary needs and preferences. Whether it's high-protein snacks for those with higher caloric needs or gluten-free options for those with sensitivities, your food selection should sustain you both physically and mentally.

- **Skill Level with Survival Tools**: Equip your bag with tools that match your skill set. A high-tech water purifier is of little use if you don't know how to operate it under stress. Include tools that you are comfortable and familiar with and consider taking courses or practicing with new tools before they make it into your bag.

Periodic Review: Keeping Your Bag Relevant and Ready

The world changes, and so do your needs. Regularly reviewing the contents of your bug-out bag will ensure that it remains up-to-date with your current situation. This periodic audit should account for:

- **Seasonal Changes**: Adjust clothing and shelter options as the weather shifts, ensuring protection against the elements whether it's the height of summer or the depths of winter.

- **Changing Family Dynamics**: Growth and change in your family, such as the addition of new members or the aging of existing ones, can shift your needs. Update your bag to reflect these changes, ensuring everyone's needs are covered.

- **Advancements in Technology or Knowledge**: As new survival tools and techniques become available, evaluate your kit for potential upgrades. Stay informed about the latest in emergency preparedness to keep your bag at the cutting edge of survival technology.

Personalizing your bug-out bag transforms it from a mere collection of items into a curated toolkit designed for your survival and comfort. By taking the time to assess individual needs, tailor the contents to those needs, and regularly review and update your pack, you will ensure that your bug-out bag is not just prepared but personalized, ready to support you and your loved ones in any situation.

SUSTAINABLE PRACTICES: EMBRACING THE OFF-GRID ETHOS

Transitioning to an off-grid lifestyle is a profound commitment to living in harmony with the earth. It's about adopting practices that not only sustain you but also enrich the environment you inhabit. In this journey, sustainability isn't just a buzzword; it's the very foundation of a resilient, self-sufficient life. Here, we'll delve into the core techniques that enable a truly off-grid existence, focusing on renewable energy, waste management, and food production through permaculture.

OFF-GRID TECHNIQUES: BUILDING A SUSTAINABLE FOUNDATION

Harnessing the Sun: Solar Power Systems

The sun offers an abundant, clean source of energy, making solar power systems a cornerstone of off-grid living. Starting with a basic setup can illuminate your path to energy independence. Begin by installing a few solar panels to power essential devices and lighting. As your comfort with the technology grows, you can expand your system to accommodate larger energy needs. Real-world examples abound of homesteads that have achieved complete energy self-sufficiency through gradual investment in solar infrastructure, demonstrating the viability and efficiency of solar power in various climates and conditions.

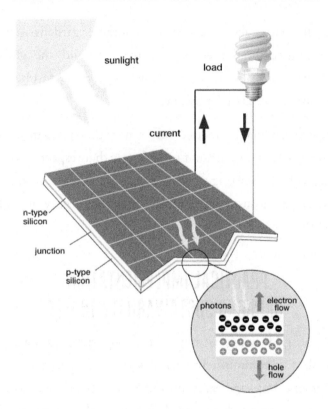

Waste Not, Want Not: Sustainable Waste Management

Effective waste management is crucial in minimizing your environmental footprint. Composting organic waste transforms leftovers into nutrient-rich soil, perfect for gardening, while recycling programs can reduce landfill contributions. Innovative approaches, such as constructing wetland systems for greywater treatment, illustrate how off-gridders can turn waste into resources, supporting both their homestead and the surrounding ecosystem.

Cultivating Abundance: Establishing a Permaculture Garden

Permaculture goes beyond gardening; it's a philosophy for working with, rather than against, nature. By mimicking natural ecosystems, permaculture gardens produce a diverse range of foods and medicinal plants, while fostering biodiversity and soil health.

Starting small—with a few raised beds or container gardens—allows you to experiment and learn. Success stories from around the globe provide inspiration and practical insights, showing how even modest beginnings can evolve into lush, productive landscapes.

Adopting these sustainable practices offers more than just a means to live off-grid; they represent a commitment to a lifestyle that respects and replenishes the natural world. By starting small and scaling up, you can transform your off-grid dream into a sustainable reality, creating a legacy of resilience and harmony with the earth.

PRACTICAL IMPLEMENTATION: BRINGING SUSTAINABILITY TO LIFE

Embracing off-grid living is not just about stepping away from the conventional; it's a commitment to a lifestyle that respects and rejuvenates the environment. Implementing sustainable practices is the cornerstone of this philosophy, turning ideals into daily actions that ensure a harmonious existence with nature. This section provides a practical guide to making sustainability not just a concept but a lived reality.

Step-by-Step to Sustainability

1. Setting Up Solar Power:

- **Assessment:** Start by evaluating your energy needs. This assessment will guide the size and type of solar setup required.
- **Installation:** Begin with a small, manageable system. A basic setup might include a few solar panels, a charge controller, a battery bank, and an inverter.
- **Expansion:** As you become more comfortable and your energy needs grow, you can add more panels and batteries accordingly.

2. Sustainable Waste Management:

- **Composting:** Separate organic waste for composting. This not only reduces landfill contributions but also enriches your garden soil.
- **Recycling and Reusing:** Implement systems for recycling materials. Explore creative ways to reuse items to minimize waste.
- **Wastewater:** Consider installing a greywater system for watering plants or flushing toilets, significantly reducing water usage.

3. Establishing a Permaculture Garden:

- **Design:** Begin with a design that mimics natural ecosystems, focusing on plants that are native to your area and beneficial to each other.
- **Implementation:** Start small, perhaps with a vegetable patch or herb garden, and expand as you gain experience and confidence.
- **Maintenance:** Utilize natural pest management and composting to maintain your garden, keeping it organic and sustainable.

Navigating Challenges and Celebrating Benefits

Implementing these techniques can come with its share of challenges, from the initial learning curve of managing a solar setup to the patience required in establishing a thriving permaculture garden. However, the benefits far outweigh these hurdles. Not only do these practices lead to a significantly reduced carbon footprint and lower living costs, but they also foster a deeper connection with the environment and a sense of achievement in self-sufficiency. Studies and real-world examples abound of individuals and communities flourishing through the adoption of sustainable off-grid practices. These stories not only inspire but also illustrate the tangible benefits of a lifestyle in harmony with nature.

Now is the moment to step into the life you envision—one of resilience, independence, and harmony with the natural world. Let the knowledge and strategies shared in this chapter be your guide as you lay the foundation for a sustainable, self-sufficient existence. Remember, each action you take, no matter how small, is a stride toward a more empowered and peaceful life. As you turn this page, anticipate delving deeper into the essence of off-grid living. The journey ahead is rich with exploration into advanced survival skills, the art of community building within an off-grid context, and comprehensive guides on harnessing renewable energy sources. This next chapter promises to elevate your preparedness, weaving together the threads of knowledge and action into a tapestry of true off-grid mastery.

Going off-grid? These 10 indispensable bonuses will help, including an expanded medical handbook and sustainable living practices. Don't forget to scan this QR code for your bonuses! Alternatively, you can visit this website:

https://rebrand.ly/scan-for-your-bonuses

PART 4:
SELF-SUFFICIENCY SKILLS

CHAPTER 10
THE ULTIMATE PREPPER'S COOKBOOK

"Food is not rational. Food is culture, habit, craving and identity."
— *Jonathan Safran Foer*

In the shadows of uncertainty and the great outdoors, where resources are a treasure, the art of cooking transcends mere survival. It becomes a celebration of resilience, a testament to the human spirit's ability to find joy and comfort in the simplest of meals. This chapter unfolds the secrets to crafting gourmet experiences with the bare minimum, turning the concept of survival cooking into an exploration of culinary creativity and sustenance.

Navigating through this journey, we'll delve into the world of gourmet survival recipes, discovering how to elevate common stockpiled ingredients into meals that nourish the body and soul. With a nod to the wisdom of experts and the annals of survival history, we'll unlock the versatility of staples, like lentils and rice, transforming them into dishes that remind us of the richness of life beyond the basics.

Equally, we'll step outside the conventional kitchen, embracing electricity-free cooking methods that reconnect us with the primal essence of fire and sun. Through solar ovens, wood-fired stoves, and ingenious makeshift grills, you can learn not just to cook but to commune with the elements, crafting meals that are as satisfying to prepare as they are to consume.

Preserving and fermenting foods emerged not just as techniques for extending shelf life but also to enrich flavors and nourish the gut. We'll traverse the science and art behind fermentation, guided by the insights of pioneers in the field, to unveil a world of tastes and textures that can elevate the prepper's pantry from mere necessity to a culinary treasure trove.

Embark on this culinary odyssey not just as a means to endure but to thrive and delight in the flavors of life, even in its most challenging moments. Let this chapter be your guide to rediscovering the joy of cooking as an essential thread in the fabric of survival, weaving together techniques, tastes, and traditions that feed more than just the body.

GOURMET SURVIVAL RECIPES: FEASTING BEYOND SURVIVAL

The essence of survival is not just to endure but to thrive, and what better way to embrace life than through the joy of eating well? Nutrition and morale go hand in hand, especially in survival situations where the right meal can uplift spirits and fortify the body against challenges. This segment of the chapter celebrates the art of creating gourmet meals that not only nourish but also delight, utilizing the staples that grace the shelves of any well-prepared pantry.

NUTRITIOUS AND ENJOYABLE: CRAFTING CULINARY DELIGHTS

In survival, each ingredient holds the potential for greatness. Lentils, rice, canned vegetables, and dried meats, when combined with a dash of creativity and a pinch of culinary skill, can transform into dishes that comfort, satisfy, and energize. Nutrition experts like Johnson (2020) remind us that proteins like lentils are not only a powerhouse of nutrients but also incredibly versatile, able to take on flavors and textures that elevate them beyond their humble origins.

Consider the simplicity and richness of a lentil stew, simmered with herbs and spices, perhaps with a bit of preserved meat for depth and complexity. Or imagine transforming rice into a savory pilaf, enriched with dried fruits and nuts, offering a perfect blend of carbohydrates for energy and sweetness to lift the spirits.

From Basics to Gourmet: A Culinary Transformation

The transition from basic ingredients to gourmet meals in a survival scenario is not just about the recipes but the approach. Viewing each component with respect and understanding its nutritional value and flavor profile allows for inventive combinations that please the palate and nourish the body.

1. **Bean and Grain Symphony:** A melody of beans mixed with grains can create complete proteins, essential for maintaining muscle strength and health. Spices, available from any pantry, add not just flavor but also antioxidants and medicinal benefits.
2. **Vegetable Medley:** Canned or dehydrated vegetables can be rehydrated and sautéed with garlic and herbs to create a side dish that rivals any fresh garden produce in taste and satisfaction.
3. **Meat Reimagined:** Preserved meats, whether canned, dried, or smoked, can be shredded and added to soups, stews, or even salads, providing essential fats and flavors that make each meal more fulfilling.

Historical uses of these ingredients in survival contexts underscore their value and versatility. During times of scarcity, cultures around the world have leveraged similar staples to create dishes that not only sustained the body but also the community spirit. These recipes are a testament to the resilience and ingenuity of the human spirit, proving that even in the most challenging times, we can find comfort, joy, and satisfaction in the meals we prepare.

By embracing the principles of nutrition and enjoyment in our survival cooking, you can elevate the act of eating from mere sustenance to a celebration of life and resilience. Through these gourmet survival recipes, you can not only feed your body but also nurture your soul, finding joy and contentment in every bite.

VERSATILITY IN INGREDIENTS: A SURVIVAL COOK'S CREATIVITY

The cornerstone of survival cooking lies not in the abundance of ingredients but in the ability to adapt, innovate, and tailor recipes to meet ever-changing circumstances. Whether dictated by dietary restrictions, fluctuating availability of resources, or a desire for culinary diversity, the versatility of ingredients in your pantry can elevate your meals from routine to remarkable. This section explores how to master the art of substitution and variation, ensuring every dish can be customized without sacrificing taste or nutritional integrity.

Mastering the Art of Substitution

Adaptability in the kitchen becomes crucial when resources are limited. Learning to substitute ingredients effectively allows for a broader range of culinary possibilities, ensuring that meals remain interesting and nutritious, even under constraints. Studies on ingredient substitutes provide a wealth of knowledge for survival cooks looking to maintain a balanced diet when favorites may not be available.

1. **Protein Switches:** Beans, lentils, and canned meats can often substitute for one another in recipes. Each offers unique nutritional benefits, from fiber in beans to high protein in meats. Smith et al. (2022) highlighted the nutritional equivalence of various protein sources, ensuring that meals can remain balanced, regardless of the specific protein used.

2. **Vegetable Variations:** Depending on what's available, vegetables can be swapped in most recipes. Root vegetables offer similar textures and can replace each other, while leafy greens can be varied based on preference and availability, each bringing its own set of vitamins and minerals to the table.

3. **Grain Alternatives:** Rice, quinoa, barley, and even rolled oats can serve as interchangeable bases for many dishes. Their versatility is not just in texture or taste but in the energy and essential nutrients they provide. For those with gluten sensitivities, quinoa and oats can offer a gluten-free alternative without compromising on the dish's essence.

Tips for Successful Ingredient Substitution

- **Taste and Texture:** Consider both the flavor and texture of substitutes to ensure they complement the dish. Experimentation and adjustment are key.

- **Nutritional Value:** Choose substitutes that offer similar nutritional profiles to maintain the meal's health benefits. For instance, replacing a carbohydrate with another of similar glycemic index will ensure energy levels remain consistent.

- **Availability and Sustainability:** Prioritize substitutions based on what's readily available and sustainable within your environment. This not only reduces waste but also encourages a deeper connection with the food you consume.

The ability to creatively substitute and vary ingredients is a testament to the resilience and ingenuity of those committed to off-grid living and survival preparedness. By embracing the principles of versatility, survival cooks can ensure that every meal is not just a means to an end but a celebration of adaptability and flavor.

ELECTRICITY-FREE COOKING: EMBRACING THE FLAMES AND THE SUN

In the heart of off-grid living and survival preparedness, the ability to cook without electricity is not just a skill; it's a liberation from the grid's constraints, a return to the basics of human ingenuity. This chapter illuminates the path to self-sufficiency through the use of ancient methods modernized for today's survivalists. From the radiant heat of the sun to the primal flame of fire, we'll explore cooking techniques that can connect us more deeply to the elements and our ancestors.

ALTERNATIVE COOKING METHODS: HARNESSING NATURAL ENERGY

Solar Ovens: The sun, our most reliable and abundant energy source, offers a cooking method that requires patience but provides efficiency and safety. Solar ovens concentrate sunlight to a focal point where it can cook food over several hours without fuel or flame. Setting up a solar oven involves choosing a sunny location and angling the reflectors to maximize sunlight concentration. Experts in solar cooking emphasize the importance of timing and temperature control, advising beginners to start with simple recipes, like baked goods or slow-cooked stews.

Wood-Fired Stoves: For those who crave the crackle of a flame, wood-fired stoves provide a robust solution for cooking. The key to their efficient use lies in understanding wood types and fire management techniques that produce consistent heat. Historical precedents show that hardwoods, like oak and ash, offer longer burn times, making them ideal for sustained cooking. The art of managing a wood-fired stove involves maintaining a bed of coals that can be stoked or dampened to control the temperature, allowing for a range of cooking methods from boiling to baking.

Makeshift Grills: Sometimes, survival cooking requires improvisation. Makeshift grills can be assembled from bricks or stones or even dug into the

150

earth, with a grate placed over the top for grilling. These rudimentary setups harness direct flame for cooking, offering the added benefit of smoke for flavoring. Seasoned grillers recommend using natural, non-treated wood to avoid chemicals in the food, emphasizing the importance of wind direction in controlling smoke and flame.

Each of these electricity-free cooking methods not only ensures the ability to prepare meals under any circumstances but also deepens our connection to the natural world and our resourcefulness. By mastering these techniques, one can embrace a form of culinary independence that modern conveniences cannot replicate, ensuring that even in the most challenging situations, the act of cooking remains a source of joy and satisfaction.

PRACTICAL DEMONSTRATIONS: BRINGING HEAT TO THE OFF-GRID KITCHEN

In the heart of the wilderness or the quiet of a blackout, the ability to conjure warmth and nourishment without electricity is both an art and a science. This section transitions from theory to practice, providing a hands-on guide to mastering alternative cooking methods. Through step-by-step instructions, you'll learn not only to cook but to cook well, using the tools and techniques suited for a life untethered from the grid. Safety, efficiency, and flavor are the cornerstones of this culinary adventure.

Solar Oven Delights

Cooking with the sun is both an efficient and eco-friendly method. Here's how to prepare a simple, hearty stew using a solar oven:

1. **Preparation:** Gather your ingredients—diced vegetables, beans, and any available meats. Season well with salt, pepper, and any herbs you have on hand.

2. **Assembly:** Place the ingredients in a dark, lidded pot to absorb more solar energy. If available, a little broth or water can help prevent burning and add flavor.

3. **Cooking:** Position your solar oven toward the sun, with the pot inside. Cooking times will vary based on weather conditions, but a bright day can have your stew ready in 2-4 hours.

4. **Safety Tip:** Use gloves when handling the pot because it can get surprisingly hot. Ensure the food reaches a safe temperature to kill any harmful bacteria.

Wood-Fired Mastery

The primal act of cooking over an open fire brings flavor and warmth to any meal. Here's a method for grilling fish to perfection:

1. **Setup:** If you don't have a grill, create a makeshift one with green branches or a flat stone placed over a bed of coals.

2. **Preparation:** Season your fish with herbs and a little lemon, if available. Wrap it in leaves or foil to retain moisture.

3. **Cooking:** Place the fish over your coals, not open flames, to avoid charring. Cook each side for about 4-6 minutes, depending on thickness.

4. **Safety Tip:** Ensure the fish is thoroughly cooked by checking for an opaque and flaky texture. Keep a safe distance from the flames to avoid burns.

Efficient Energy Use

Whichever method you choose, maximizing energy efficiency will ensure you get the most out of your cooking experience:

- **Pre-heat:** Gather heat before adding your pot or grill to shorten cooking times.
- **Minimize opening:** Every time you open a solar oven or uncover your food, heat escapes. Check only when necessary.
- **Utilize residual heat:** Even after removing from the heat source, your food continues cooking. Plan for this to avoid overcooking.

By embracing these practical demonstrations of electricity-free cooking, you can not only broaden your culinary skills but also prepare yourself for a life of independence and resourcefulness. Cooking can become more than a means to an end—it can transform into an enjoyable, rewarding part of your daily routine.

PRESERVING AND FERMENTING: TIME-HONORED TRADITIONS FOR THE MODERN PREPPER

In the quest for self-sufficiency, the ancient arts of preserving and fermenting foods emerge as essential skills, bridging past and present to fortify our future.

These techniques not only extend the shelf life of precious provisions but also unlock deeper nutritional value and introduce a palette of flavors that can enrich the prepper's table. This section delves into the sophisticated yet accessible world of food preservation, illuminating the science behind fermentation and offering step-by-step guides to harnessing this natural process.

ADVANCED PRESERVATION TECHNIQUES: CULTIVATING CULINARY RESILIENCE

Fermentation stands as a cornerstone of food preservation, a biological wonder that transforms simple ingredients into long-lasting, flavor-enhanced foods. Through the action of beneficial bacteria, yeasts, and molds, fermentation can safely extend the edible life of fruits, vegetables, meats, and dairy, all while boosting their nutritional profile.

1. **Sauerkraut Magic:** Begin with cabbage, salt, and a clean jar. Shred the cabbage, mix it with salt, and pack it tightly into the jar. The salt draws out water, creating an anaerobic environment where lactobacillus (friendly bacteria) thrives. Seal and store at room temperature. In a week, you'll have sauerkraut.
2. **Kefir Creation:** Combine kefir grains with milk in a jar, cover with a cloth, and let it sit. The grains ferment the milk, turning it into a probiotic-rich kefir in about 24 hours.
3. **Pickling Vegetables:** Vegetables, water, salt, and optional spices, like dill or garlic, are all you need. The brine solution creates the perfect conditions for fermentation, resulting in tangy, crunchy pickles.

The Science of Safety and Flavor

Understanding the scientific principles of fermentation is crucial for safety and success. Fermentation primarily relies on creating conditions conducive to beneficial microorganisms while inhibiting harmful ones. According to

Jones and Lee (2021), the right balance of salt, temperature, and time not only ensures the safe preservation of foods but also enhances their flavors and nutritional content.

Fermentation transforms the mundane into the extraordinary, bringing a new dimension to off-grid living and preparedness. Whether it's the tang of homemade yogurt, the crunch of freshly fermented pickles, or the complex flavors of aged cheese, these preserved delights add richness to meals and a sense of accomplishment to the prepper's journey. By embracing these advanced preservation techniques, we can connect with traditions that have sustained humanity for millennia, armed with the knowledge and skills to nourish ourselves and our loved ones in any circumstance.

HEALTH AND SAFETY: THE CORNERSTONES OF SUCCESSFUL FERMENTATION

Navigating the world of food preservation, particularly fermentation, requires a keen understanding of health and safety practices. While fermenting foods can opens up a treasure trove of flavors and nutritional benefits, ensuring the process is done safely is paramount. This part of the chapter focuses on essential safety guidelines that safeguard both the quality of your fermented products and the health of those who enjoy them.

Safety First: Guidelines for Successful Fermentation

When fermenting foods, creating conditions that favor beneficial bacteria while inhibiting harmful pathogens is crucial. Here are key safety measures to observe:

- **Cleanliness:** Begin with a clean workspace, utensils, and containers. Sterilization isn't necessary for fermentation, but minimizing unwanted bacteria is essential.

- **Salt Concentration:** Follow recommended salt concentrations closely. Salt inhibits harmful bacteria, ensuring that only beneficial microbes thrive.
- **Temperature:** Keep fermenting foods at the correct temperature. Too warm, and harmful bacteria may proliferate; too cool, and fermentation may not occur effectively.
- **pH Levels:** Monitor the acidity of your ferments. A pH of 4.6 or lower is generally safe, as it prevents the growth of botulism bacteria.

Recognizing and Avoiding Hazards

Knowing what signs indicate successful fermentation versus potential spoilage is vital. Healthy ferments often have a tart, tangy smell and taste. Any signs of mold, unusual odors, or colors suggest contamination and should be discarded.

Nutritional Powerhouses

Fermented foods aren't just safe; they're superfoods. Fermentation can increase the availability of vitamins and minerals, making them more digestible and adding beneficial probiotics that support gut health. Food safety authorities and nutritional research, such as findings by the World Health Organization, have documented the enhanced nutritional profiles of fermented foods and their role in a balanced diet. Adhering to these safety guidelines ensures that your foray into the world of fermented and preserved foods is not only creative and delicious but also secure and nutritious. With each batch of sauerkraut, kefir, or pickles, you're not just extending the shelf life of your food; you're enriching your diet with flavors and nutrients that support overall well-being.

Embrace the journey ahead with spatula in hand and spirit undaunted. The recipes and techniques outlined in this chapter are not just instructions; they are an invitation to explore the resilience and richness of life, even in the most challenging times. Good food is more than sustenance; it is comfort, culture, and a critical component of preparedness that nourishes both body and soul. Now is the time to begin experimenting with the flavors and fermentations that can transform your survival stockpile into a gourmet pantry. Let each meal be a testament to your adaptability and creativity, proving that even in the harshest conditions, the human spirit can thrive. Step forward with confidence, knowing that each page turned is a step closer to mastering the art of resilience. The journey of preparedness is ongoing, and the next chapter awaits to further inspire and empower you on this path.

Thank you for selecting "The Complete Prepper's Survival Bible." If you enjoyed this Ultimate Prepper's Cookbook chapter, you'll appreciate "The Ultimate Prepper's Pantry Cookbook," one of the 10 bonuses you'll receive when you scan the QR code above. Alternatively, you can visit this website: https://rebrand.ly/scan-for-your-bonuses

CHAPTER 11
MAXIMIZING YOUR PREPPER GARDEN

"The greatest fine art of the future will be the making of a comfortable living from a small piece of land."
— *Abraham Lincoln*

Gardening, at its core, is an act of resilience. It's the bridge between the raw hand of nature and the nurturing touch of human care, a dance of patience and reward that spans seasons and generations. This chapter dives into the heart of prepper gardening, exploring techniques that push the boundaries of traditional horticulture to ensure food security and sustainability. From the ingenious adaptation of year-round gardening strategies to the futuristic realms of soilless agriculture, we'll chart a course toward complete self-sufficiency.

Innovation blooms at every turn, with cold frames and mulches extending the growing season beyond the confines of summer and small-scale greenhouses capturing the sun's bounty. The domestication of the wilderness into verdant plots of nourishment speaks to the prepper's ultimate goal: resilience through self-reliance.

The journey then ventures into the soilless wonders of hydroponics and aquaponics, where water becomes the lifeblood of lush harvests, proving that even the smallest of spaces can yield abundant crops. These methods not only conserve precious resources but also offer a blueprint for the future of sustainable agriculture.

Preservation of biodiversity through seed saving emerges as a vital thread in the fabric of prepper gardening, ensuring that the genetic tapestry of our food sources remains rich and varied. This practice stands as a testament to the importance of maintaining a link to the past as we forge ahead into uncertain futures.

YEAR-ROUND GARDENING: MASTERY OVER THE SEASONS

The ability to harvest fresh produce throughout the year is a game-changer for any prepper looking to achieve true food independence. Extending the growing season beyond the traditional boundaries of spring and summer not only ensures a steady supply of food but also diversifies your diet and enhances your garden's resilience. By adopting a few key techniques, you can protect your plants from the chill of winter and the scorch of summer, turning your garden into a year-round bounty.

EXTENDING THE GROWING SEASON: TECHNIQUES FOR TRIUMPH

Cold Frames: Natural Warmth for Tender Plants

Cold frames, essentially miniature greenhouses, harness the sun's warmth to create microclimates suitable for plant growth even when the air outside is cold. Constructed from transparent materials atop a weather-resistant frame, they capture solar energy, significantly raising the internal temperature. Jones (2022) emphasizes their effectiveness, noting that cold frames can extend the growing season by several weeks at both the beginning and end, allowing for the cultivation of greens like spinach and lettuce well into colder months.

Mulches: The Protector of Soil's Secrets

Mulch is more than just a garden accessory; it's a vital tool for temperature and moisture regulation. By applying a layer of organic material, such as straw or wood chips, over the soil, you can create an insulation layer that protects roots from temperature extremes and retains moisture.

160

Row Covers: A Shield Against the Elements

Lightweight yet effective, row covers consist of a breathable fabric draped over plants, protecting them from frost, wind, and even pests. They allow light and water to reach the plants while insulating them from the cold. The versatility of row covers means they can be used from early spring through late fall, extending the growing season for a wide variety of crops.

Embracing these techniques allows you to defy the traditional gardening calendar, providing fresh produce even when snow blankets the ground. With each method offering a unique approach to season extension, combining them can create a garden that is not only productive all year round but also a testament to the resilience and ingenuity of its keeper.

GREENHOUSE CONSTRUCTION: HARNESSING SUNLIGHT FOR SUSTENANCE

The magic of a greenhouse lies in its ability to bend the rules of nature, creating a sanctuary of warmth and light that defies the seasons. This marvel of ingenuity allows gardeners to cultivate a bounty of fruits, vegetables, and herbs year-round, turning the dream of self-sufficient living into a tangible reality. Within the transparent walls of a greenhouse, every day is a day of growth, regardless of the snow, rain, or frost outside.

Blueprint for Growth: Building Your Greenhouse

Constructing a small-scale greenhouse is a journey that begins with envisioning your space as a nexus of productivity and sustainability. Here's a guide to laying the foundations of your year-round garden:

1. **Material Selection:** Durability meets clarity in the choice of materials for your greenhouse. Polycarbonate panels offer an excellent balance, providing the insulation needed to keep warmth in while allowing ample sunlight to reach your plants.
2. **Design Considerations:** The design of your greenhouse should cater to the specific needs of your garden while optimizing environmental

conditions. A peaked roof not only prevents snow accumulation but also maximizes headspace for taller plants. Ventilation is crucial; ensure your design includes adjustable windows or vents to regulate temperature and humidity.

3. **Location Selection:** Positioning is everything. Choose a location that receives maximum sunlight throughout the day. A south-facing orientation is ideal in the northern hemisphere, as it captures the sun's rays most effectively. Additionally, consider proximity to water sources and protection from strong winds.

4. **Maximizing Solar Gain:** The secret to a flourishing greenhouse lies in its ability to harness solar energy. Smith et al. (2020) has provided a comprehensive guide on maximizing solar gain through strategic design and placement. Incorporating thermal mass, such as water barrels or concrete floors, can absorb heat during the day and release it at night, maintaining a consistent temperature.

By embarking on the construction of your greenhouse, you're not just building a structure; you're creating an ecosystem. This space will serve as a testament to what can be achieved when human ingenuity and the forces of nature converge. With each seed planted and each harvest gathered, your greenhouse becomes a beacon of resilience, offering a year-round supply of fresh, nutritious produce right at your doorstep.

INDOOR GARDENING: A CONTROLLED OASIS

Transforming the interior of a home into a lush, productive garden space unveils the remarkable potential of indoor gardening. This innovative approach brings the art of cultivation right into the heart of your living space, offering unparalleled control over the growing environment. Here, every element—from light to temperature to humidity—can be fine-tuned to create optimal conditions for plant growth, all year round.

CRAFTING YOUR INDOOR GARDEN

The foundation of a successful indoor garden lies in understanding and implementing the right setup. With the advancement of technology and gardening techniques, establishing a green oasis indoors has never been more accessible or efficient.

1. **Light the Way:** Plants' primary need is light, and in indoor settings, the sun's rays often need help to suffice. Enter the world of grow lights. LED grow lights have revolutionized indoor gardening, offering a spectrum of light that mimics natural sunlight, promoting photosynthesis without excessive heat. Johnson (2021) underscores the effectiveness of LED grow lights in indoor horticulture, highlighting their energy efficiency and the broad range of light wavelengths they provide, catering to the needs of various plant types.

2. **Choosing Containers:** Selecting the right containers is next. Pots with good drainage are essential to prevent waterlogging. Consider the size of the mature plant to ensure the container can accommodate its growth.

3. **Mastering the Environment:** Beyond light and space, managing temperature and humidity levels is crucial. Most indoor plants thrive in temperatures between 65°F and 75°F. Humidity can be more challenging, especially in dry climates or winter months. Simple solutions like a tray of water near your plants or a small room humidifier can help maintain a more tropical atmosphere.

By embracing indoor gardening, you're not just bringing nature indoors; you're setting the stage for a year-round harvest that is both rewarding and sustainable. This controlled oasis will not only contribute to your household's self-sufficiency but will also enhance your living space with beauty and vitality. As you delve into this green endeavor, remember that the journey is often as fulfilling as the harvest.

STEP-BY-STEP SETUP: CULTIVATING SUCCESS IN SOILLESS GARDENS

Embarking on the journey of soilless gardening can open a new chapter in the cultivation narrative, where water becomes the new earth, and symbiosis paves the way for abundance. This guide seeks to demystify the process of setting up both hydroponic and aquaponic systems, ensuring even beginners can confidently grow their gardens without a speck of soil.

Hydroponics Setup: Your Foundation for Flourishing Plants

Creating a hydroponic garden begins with assembling a few key components:

1. **Choose Your System:** Start with a simple setup like a deep-water culture (DWC) or a nutrient film technique (NFT) system. Each has its benefits, with DWC being great for beginners due to its simplicity.

2. **Gathering Supplies:** You'll need a container or reservoir, an air pump (for oxygenating the water), net pots, and a nutrient solution. Don't forget grow lights if you're setting up indoors.

3. **Mixing Nutrients:** Nutrient solutions are the lifeline of your hydroponic garden. Green (2020) provides an essential guide on mixing the right formulations for different plants, such as lettuce, emphasizing the importance of balancing macronutrients and micronutrients to mimic natural soil conditions.

4. **Planting:** Start with seedlings that have been germinated in rockwool cubes or similar growing mediums. Transfer them to net pots, ensuring the roots can reach the nutrient solution.

5. **Maintenance:** Monitor pH levels and nutrient concentrations regularly, adjusting as needed to keep your plants thriving.

Aquaponics Setup: Harmonizing Fish and Plants

An aquaponic system takes the hydroponic principle and enriches it with the natural fertilizer fish provide, creating a closed-loop ecosystem.

1. **System Design:** A basic aquaponics system combines a fish tank with grow beds. The water from the fish tank is pumped into the grow beds, where plants filter and clean it before it returns to the fish, rich in oxygen.

2. **Choosing Fish and Plants:** Select fish species well-suited for your climate and tank conditions. Davis (2019) highlights the efficiency of using tilapia or carp, given their hardiness and growth rate. Leafy greens, like lettuce, herbs, and other non-root vegetables, thrive in aquaponic systems.

3. **Balancing the Ecosystem:** Achieving the right balance between the number of fish and the plant load is crucial. Start small to avoid overwhelming the system. Monitor water quality closely, paying attention to ammonia, nitrite, and nitrate levels, to ensure a healthy environment for both fish and plants.

4. **Ongoing Care:** Feed your fish quality food, maintain water levels, and check the pH and nutrient levels regularly. As the system matures, it will require less intervention, moving closer to a self-sustaining model.

By following these foundational steps, enthusiasts can explore the rewarding process of soilless gardening. Whether through the pure plant focus of hydroponics or the dynamic fish-plant interplay of aquaponics, these methods offer efficient, sustainable pathways to food production, unlocking new possibilities for gardeners everywhere.

SEED SAVING: PRESERVING THE PAST, PLANTING THE FUTURE

Seed saving is more than a practice; it's a passage through time, a means of capturing the essence of today's harvest to sow the seeds of tomorrow's bounty. This timeless tradition is crucial for maintaining biodiversity, ensuring garden resilience, and safeguarding our agricultural heritage. By focusing on heirloom varieties, gardeners can play a vital role in preserving genetic diversity, ensuring that the flavors, stories, and resilience embedded in these seeds are passed on to future generations.

FUNDAMENTALS OF SEED SAVING: A LEGACY IN EVERY SEED

The journey of seed saving begins with the careful selection of plants. Heirloom varieties, known for their unique characteristics and adaptability, are particularly prized. These seeds carry the tales of cultures, climates, and the hands that have tended them through the ages. The process of saving these treasures involves a few key steps:

1. **Selection:** Look for healthy, robust plants that exhibit desirable traits, such as disease resistance, productivity, and superior taste. These are the specimens most worthy of passing on their legacy.
2. **Harvesting:** Timing is crucial. Seeds should be collected at the peak of maturity, which varies among plant species. For many vegetables and fruits, this means waiting until the end of the season, when fruits have ripened and seeds have fully developed.
3. **Preparing:** Once harvested, seeds need to be cleaned and dried. For dry-seeded crops (like beans and lettuce), this means simply removing the

seeds and spreading them out to dry. Wet-seeded crops (such as tomatoes and cucumbers) require a bit more effort, involving fermentation to remove the gelatinous coating, followed by drying.

4. **Storing:** Proper storage is vital to preserving the viability of seeds. Once dry, seeds should be kept in a cool, dry place, ideally labeled with the variety and date of harvest. This careful stewardship ensures that each seed's potential to germinate and grow into a healthy plant is preserved.

By engaging in the practice of seed saving, gardeners become custodians of biodiversity and contributors to a sustainable future. Each saved seed is a vote for diversity, a step toward resilience, and a bridge to the past, allowing us to grow not just gardens but also traditions and stories that enrich our world.

COLLECTION AND STORAGE TECHNIQUES: ENSURING SEED VIABILITY FOR GENERATIONS

Securing the future of your garden lies not only in the soil but in the careful collection and preservation of seeds. This vital process ensures that the genetic diversity and the hard-earned success of this year's harvest can be carried forward into the next season and beyond. Properly collecting, cleaning, drying, and storing seeds are the cornerstones of a sustainable gardening practice, enabling each gardener to become a steward of both their food supply and the broader botanical heritage.

Ensuring Seeds' Legacy Through Meticulous Collection and Storage

1. **Collection:** The first step in securing a seed's future is its careful selection and collection from the healthiest and most productive plants in your garden. This not only guarantees quality but also the perpetuation of desirable traits.

2. **Cleaning:** Once collected, seeds need to be freed from their husks, pods, or flesh. This process varies from simple shaking and winnowing for dry

seeds to fermentation for seeds encased in fleshy fruits, a step that helps remove the pulp and inhibits seed-borne diseases.

3. **Drying:** Proper drying is crucial. Seeds should be spread out in a warm, airy space away from direct sunlight. This phase ensures that seeds are sufficiently dry to prevent mold growth during storage, a key factor in maintaining their viability.

4. **Storing:** It is important to store seeds in a cool, dry place. Moisture and heat are the enemies of seed viability. Using airtight containers, such as glass jars with silica gel packets to absorb moisture, can significantly extend a seed's life. Labeling each container with the seed type, variety, and date of harvest ensures easy organization and retrieval.

5. **Organizing:** Developing a system for organizing your seeds can streamline your planting process in subsequent seasons. Whether arranged by plant family, season of planting, or any other system that suits your gardening style, efficient organization saves time and ensures that no seed goes to waste.

By adhering to these best practices for seed collection and storage, gardeners can ensure the longevity and productivity of their seed stock, creating a self-renewing resource that supports both the current garden and those of future seasons. This cycle of renewal is a testament to the resilience and foresight inherent in the gardening spirit, embodying a commitment to sustainability and self-sufficiency.

Now is the time to put soil under your fingernails and the sun on your back, to take the seeds of knowledge from this chapter and plant them in the fertile ground of your garden. Let the empowerment of growing your food sustainably be your guide, transforming your relationship with the earth beneath your feet and the food on your plate. This is not just gardening; it's a step towards resilience, a declaration of independence, and a commitment to a sustainable future.

As you turn the page, anticipate an exploration of advanced water collection and irrigation techniques, essential skills for nurturing your garden and ensuring its bounty. Imagine building a community of self-reliant gardeners, where shared knowledge strengthens the roots of sustainability. The journey ahead is rich with possibility, each chapter building on the last, guiding you deeper into the art and science of preparedness. Join us as we continue to cultivate not just gardens but a way of life grounded in self-sufficiency and harmony with nature.

Unlock valuable prepping information with 10 guides including a Seed Saving and Gardening Guide. Use your smartphone to scan the QR code above to access your bonuses. Alternatively, you can visit this website: https://rebrand.ly/scan-for-your-bonuses

CHAPTER 12
EXPERT TRAPPING AND HUNTING

"Nature is not a place to visit. It is home."

— *Gary Snyder*

Trapping, hunting, and fishing aren't just about survival; they're about connecting deeply with the earth and living in a way that honors the balance of nature. This chapter delves into the vital skills you need to thrive in the wild, focusing on doing so responsibly and sustainably. We'll explore the art of crafting and setting traps, the ethics of hunting in harmony with nature, and the wisdom of fishing with an awareness of the ecosystem. These aren't just techniques; they're pathways to a life of self-sufficiency and respect for the natural world.

Let's start with advanced trapping. Knowing how to set a trap that catches food effectively while minimizing harm to wildlife is a cornerstone of ethical wilderness living. We'll walk through various traps, from simple snares to intricate figure-four deadfalls, each designed for specific game. It's not about catching the most or the biggest; it's about sustainability and need. We adhere to the wisdom of those who've walked these paths before us, respecting the laws of nature and man alike.

Moving on to hunting, it's more than aiming and shooting. It's about understanding the environment, the behavior of your prey, and the impact of your actions. Sustainable hunting means taking only what you need, choosing your targets wisely, and ensuring that you're contributing to the health of wildlife populations rather than detracting from them. It's about skill, patience, and respect.

Fishing, too, requires more than a line and a hook. It's about knowing the waters, the fish, and the delicate balance of aquatic ecosystems. We'll cover how to improvise gear from what nature offers and employ techniques that increase your chances of a successful catch without depleting local fish stocks.

ADVANCED TRAPPING TECHNIQUES

Mastering the craft of trapping is akin to learning the language of the wild. Each trap you set is a conversation with nature, a demonstration of your understanding and respect for the creatures you aim to catch. The design of each trap reflects its purpose and the specific game it targets. Let's explore the world of advanced trapping techniques, where skill, ethics, and effectiveness converge.

TRAP DESIGNS

A well-chosen trap design can mean the difference between success and failure in the wild. Whether you're after small game, like rabbits, or larger targets, the key is selecting a trap that matches your needs while minimizing harm to the animal.

Paiute Deadfall - Ideal for small game, this trap uses a trigger system that's sensitive enough to detect the slight movements of smaller animals. The efficiency of the Paiute Deadfall lies in its simplicity and the minimal effort required to set it up. Its design ensures a quick and humane capture, reducing the animal's suffering.

Figure-Four Deadfall - A classic trap known for its reliability, the Figure-Four Deadfall is versatile and can be adapted to various sizes. It's especially effective for medium-sized game. Constructed from three sticks and a heavy weight, the trap's mechanism triggers under the slightest disturbance, making it a dependable choice for survival situations.

Spring Pole Snare - For animals that are cautious and easily spooked, the Spring Pole Snare offers a solution that is both discreet and effective. Utilizing the tension from a bent sapling or branch, this trap snaps up the prey, lifting it off the ground. This method is particularly useful for capturing game without drawing attention to the trap location.

Treadle Snare - This trap is designed for larger game, using a pressure plate to trigger the snare when stepped on. Its effectiveness lies in its ability to be camouflaged within the environment, making it nearly invisible to unsuspecting animals. The Treadle Snare is a testament to the trapper's ability to understand and anticipate the movements of their target.

Basket Traps for Fish - In waterways, basket traps provide a sustainable method for capturing fish. Designed to allow smaller fish to escape while trapping larger ones, these traps are an example of ethical trapping practices that support conservation efforts.

Each of these traps requires patience and practice to master. They are tools of survival, but they also represent a trapper's commitment to ethical and sustainable practices. By choosing the right trap for the right situation, we show our respect for nature and take only what we need, ensuring that the wilderness thrives alongside us.

ETHICAL TRAPPING

Ethical trapping stands as a cornerstone of responsible wilderness survival. It embodies our commitment to the earth and its creatures, ensuring we act not as conquerors but as stewards. This section delves into the principles that guide humane trapping, illustrating how respect for wildlife, adherence to regulations, and the application of compassionate methods form the bedrock of ethical practices in the wilderness.

Humane Trapping Methods

At the heart of ethical trapping is the commitment to cause minimal distress and harm to animals. The use of non-lethal traps, as endorsed by the National Wildlife Federation in 2020, plays a crucial role in this. These traps are designed to capture without injuring, allowing for the release of unintended or non-target species unharmed. Techniques such as live-cage traps and foot-hold traps with padded jaws reflect advancements in this area, offering trappers the means to secure food or manage wildlife populations without inflicting unnecessary suffering.

Respect for Wildlife Populations

Understanding and respecting the balance of ecosystems is vital. Ethical trappers recognize their role in maintaining the health and stability of wildlife populations. This involves selecting target species carefully, considering their reproductive rates and ecological impact, and avoiding the trapping of endangered or threatened species. By aligning practices with the natural rhythms of the wilderness, trappers contribute to the preservation of biodiversity.

Legal Compliance and Conservation Efforts

Following local and national laws is not just about legal obligation; it's a demonstration of respect for collective efforts to conserve wildlife. Regulations are often in place to protect vulnerable species, ensure sustainable harvests, and safeguard ecosystems from imbalance. Ethical trappers stay informed about these laws, obtain necessary permits, and adhere to specified seasons and quotas. This legal framework supports wildlife conservation while allowing for the responsible use of natural resources.

Guidelines from Conservation Organizations

Leading conservation organizations provide valuable resources for trappers seeking to refine their practices. Guidelines like those from the National Wildlife Federation underscore the importance of using methods that align with conservation goals. These recommendations serve as a compass, guiding trappers towards techniques that minimize ecological impact and foster a respectful relationship with nature.

Ethical trapping is a practice rich in respect, knowledge, and responsibility. It reflects a deep understanding of our place in the natural world and our duty to preserve it for future generations. By embracing these principles, trappers can sustainably harvest the gifts of the wilderness, ensuring that both humanity and nature thrive together.

CONSERVATION-MINDED HUNTING

Conservation-minded hunting is an ode to the delicate dance between humankind and nature. It's a practice rooted in the deep understanding that our survival and the health of our planet are intertwined. By adopting sustainable hunting practices, we honor this connection, ensuring that the natural world thrives alongside us. Let's explore how principles of conservation and ecological balance can guide our actions in the wilderness.

SUSTAINABLE PRACTICES

Sustainable hunting goes beyond the pursuit of game; it's about nurturing the environment that sustains us all. Here are key practices that embody this ethos:

- **Selecting Appropriate Game Species**: Focus on species that are abundant and have been scientifically evaluated as sustainable options. This approach helps maintain biodiversity and ecosystem health.

- **Adherence to Quotas**: Respect the limits set by wildlife management authorities. These quotas are based on rigorous research and are designed to prevent overharvesting, ensuring species populations remain robust for generations.

- **Habitat Preservation**: The environment where game species thrive is as crucial as the hunt itself. Practices such as leaving no trace, supporting reforestation efforts, and protecting water sources contribute to the preservation of habitats.

- **Seasonal Hunting**: Aligning hunting activities with natural breeding cycles ensures populations can recover and flourish. It's a practice that respects the rhythms of nature and the lifecycle of wildlife.

- **Ethical Harvesting**: This involves taking only what you need and making every effort to ensure a quick and humane end for the game. It's a sign of respect for the animal and the ecosystem it belongs to.

- **Supporting Conservation Efforts**: Many hunters contribute to conservation programs, whether through volunteer work, financial donations, or advocacy. This partnership between hunters and conservationists has led to significant successes in wildlife management and habitat restoration.

- **Education and Mentorship**: Sharing knowledge and ethical practices with new hunters ensures that conservation-minded hunting continues. It's about passing on a legacy of respect and stewardship for the natural world.

Conservation-minded hunting is a testament to our ability to live in harmony with the earth. It's a commitment to the future, a pledge that the beauty and bounty of the wilderness will be there for all to enjoy, today and tomorrow. By adopting these practices, hunters become active participants in the conservation of our planet, proving that humanity can indeed coexist sustainably with the natural world.

SKILL DEVELOPMENT IN CONSERVATION-MINDED HUNTING

Mastering the art of hunting is more than just a pursuit of game; it's a commitment to ethical stewardship and efficiency that honors the lives we take to sustain our own. The development of hunting skills is paramount in ensuring that every action we take in the wilderness is responsible and respectful. From the silent wait to the final act of gratitude, every step is a measure of our respect for nature. Let's delve into the essential skills of tracking, field dressing, and butchering, which together ensure minimal waste and uphold the dignity of the animal.

Tracking: The Foundation of Ethical Hunting

The ability to track is the hunter's first tool in ensuring a successful and ethical hunt. It involves reading signs left by animals, understanding their habits, and moving silently through their habitat. Tracking is not just about finding game; it's about connecting with the environment and learning to read the subtle language of the wild. This skill minimizes unnecessary disturbance to wildlife and leads to more successful, direct encounters with the intended game.

Field Dressing: A Respectful Approach

Once an animal has been harvested, field dressing becomes the critical next step. This process involves removing the internal organs to prevent spoilage, ensuring the meat remains safe and consumable. Proper field dressing is a skill that reflects the hunter's respect for the game and dedication to minimizing waste. It's about efficiency and hygiene, handling the animal with care to preserve the quality of the meat for nourishment.

Butchering: Maximizing Yield with Care

The final step in honoring the life taken is butchering—carefully processing the animal into manageable cuts of meat. This stage is where skill really comes into play, as the hunter aims to maximize the yield from the animal, ensuring nothing goes to waste. Butchering requires knowledge of anatomy, sharp tools, and a steady hand. It's a process that, when done correctly, respects the animal and provides sustenance for many.

Each of these skills—tracking, field dressing, and butchering—is a testament to the hunter's role as a conservationist. They are practices steeped in the tradition of ethical hunting, ensuring that we take from the earth in ways that are sustainable and respectful. By honing these skills, hunters not only improve their efficiency but also deepen their connection to the natural world, embodying the principles of conservation-minded hunting.

SURVIVAL FISHING

Survival fishing transcends the conventional angling experience, morphing into a vital skill in the hands of those who know how to harness nature's bounty with minimal resources. This craft is not just about catching fish; it's about ingenuity, adaptability, and the deep-rooted human instinct to thrive in the natural world. Let's dive into the art of creating effective fishing gear from what the wilderness and serendipity provide.

IMPROVISED GEAR

When traditional fishing equipment isn't within reach, nature's workshop offers all the necessary materials to fashion your own. These improvised tools not only bridge the gap between need and availability but also deepen our connection with the environment by utilizing its offerings respectfully and creatively.

- **Fishing Rods from the Wild**: Look for flexible branches strong enough to withstand the tension of a caught fish yet bendy enough to give you the leverage needed. Willow, bamboo, or young green wood can serve as excellent bases for your rod. Attach a line made from plant fibers, stripped bark, or even salvaged material from your gear.

- **Crafting Hooks with Ingenuity**: Hooks can be fashioned from a multitude of materials found in the wild. Bones, thorns, and hardened wood can be carved and shaped into effective hooks. Safety pins or paper clips found among your items can also be repurposed. The key is to ensure the hook is sharp enough to penetrate the fish's mouth and shaped in a way that prevents the catch from escaping.

- **Nets from Nature's Bounty**: Constructing a net might seem daunting, but with a little creativity, it's entirely feasible. Use strong, flexible vines or strips of bark for the net's framework. For the mesh, intertwine finer materials, such as plant fibers or even hair. While this requires patience and skill, a handmade net can be a highly effective tool for securing a meal.

Survival fishing with improvised gear is a testament to human resourcefulness and the rich resources nature provides. It can teach you to observe closely, think creatively, and act resourcefully. This will not only equip you with the means to sustain yourself but also instill a profound respect for the natural world and its abundance.

FISHING TECHNIQUES

Harnessing the water's bounty for survival demands more than just patience; it requires strategy, knowledge, and adaptability. Effective fishing techniques in survival scenarios are those that consider the environment, the behavior of fish, and the resources at hand. This section unveils various methods that enhance your chances of success, tailored to different water bodies and situations, grounded in practical wisdom and proven tactics.

Still Water Fishing: In lakes and ponds, where waters run deep and calm, understanding the fish's habits is key. Early morning or late afternoon, when fish come closer to the surface to feed, are prime times for fishing. Techniques like "dapping"—where you let your bait lightly touch the water's surface to mimic an insect—can be particularly effective, as demonstrated by survival experts, like Bear Grylls in his various survival guides.

Moving Water Fishing: Rivers and streams present a different set of challenges and opportunities. Here, the strategy is to place your bait in a way that uses the current to your advantage. Positioning yourself downstream and allowing your improvised lure to drift naturally through likely fish habitats maximizes your chances of a catch. Survivalist John Wiseman suggests in "The SAS Survival Handbook" that building a small weir or dam can help direct fish to a particular spot, making them easier to catch.

Saltwater Fishing: Coastal and sea fishing demand ingenuity, especially when you're without conventional gear. Tidal pools can be rich hunting grounds during low tide, where fish might be trapped and easier to catch by hand or with a spear. For deeper waters, constructing a fish spear from straight sticks or creating a simple line and hook setup, then casting from rocks or piers, can yield good results. Ray Mears' "Outdoor Survival Handbook" offers insights into creating effective saltwater fishing gear from debris and natural materials found along the shore.

Each environment presents unique challenges, but the underlying principle remains the same: observe, adapt, and utilize what you have with as much knowledge and respect for nature as possible. With these strategies, survival fishing becomes not just a means of sustenance but a profound engagement with the natural world, offering lessons in resourcefulness, patience, and the rhythm of the waters.

As we move forward, the adventure deepens. The next chapter unfolds into the realm of advanced foraging techniques, water purification strategies, and the art of building resilient communities. These skills will not only augment your survival toolkit but also enrich your understanding of self-sufficiency. Prepare to explore these new horizons, where your journey toward complete autonomy and harmony with nature continues, opening doors to a life of preparedness, resilience, and profound connection with the earth.

Gain access to bonuses that fit perfectly with The Complete Prepper's Survival Bible. Scan this QR code with your smartphone. Alternatively, you can visit this website:

https://rebrand.ly/scan-for-your-bonuses

CHAPTER 13
THE PREPPER'S MEDICAL HANDBOOK

"Survival of the fittest is less about strength, and more about adaptability and resilience, especially when it comes to health."

— Anonymous

Preparing for survival means preparing for every aspect of life, and health is at the forefront. This chapter goes beyond the basics of first aid to explore the full spectrum of medical preparedness necessary for long-term survival. Managing chronic conditions, navigating acute medical emergencies, and incorporating alternative therapies into your health plan are critical components of a robust survival strategy. We aim to equip you with the knowledge to not only survive but thrive, maintaining health and managing medical situations with confidence in any scenario.

Chronic conditions, like diabetes and hypertension, don't vanish when society's structures crumble. They require foresight and planning, with strategies for medication storage, monitoring, and adjustment becoming vital in the absence of regular medical supervision. Imagine a world where the local pharmacy is no longer around the corner, and the guidance of the American Diabetes Association becomes your beacon, advising a three-month supply of necessary medications as a minimum standard for preparedness.

In the realm of acute emergencies, the ability to act swiftly and effectively can be the difference between life and death. From administering CPR to managing snake bites, the knowledge to handle these situations is indispensable. This knowledge is not just about following steps but understanding the why behind each action, ensuring you can adapt to the specifics of each emergency.

The chapter also embraces the potential of alternative therapies, from acupuncture to energy healing. In environments where traditional medical treatments may be scarce, these practices, grounded in centuries of tradition and supported by contemporary research, offer valuable complements to physical and mental well-being.

CHRONIC CONDITION MANAGEMENT

When the fabric of society frays, the challenges of managing chronic conditions magnify. Without the usual medical support networks, individuals with conditions like diabetes and hypertension are left to navigate their health with far fewer resources. This segment aims to arm you with strategies that ensure continuity of care, even when traditional medical facilities are out of reach.

PRACTICAL ADVICE FOR SELF-MANAGEMENT

Managing chronic conditions in uncertain times calls for a proactive and prepared approach. The key lies in understanding your condition, foreseeing potential challenges, and having a solid plan in place.

Medication Storage and Management: Stability in medication supply is a lifeline for those with chronic conditions. The American Diabetes Association emphasizes the critical nature of maintaining at least a three-month supply of essential medications. This buffer allows for unforeseen delays in access to pharmacies and healthcare providers. Additionally, it's crucial to understand the proper storage conditions required for your medications.

Monitoring Your Condition: Regular monitoring is the cornerstone of effective chronic condition management. For individuals with diabetes, regular blood sugar testing becomes even more vital in the absence of regular physician consultations. Investing in a reliable glucose meter and ample testing strips and learning to interpret the results accurately can empower you to make informed decisions about your diet, activity, and medication.

Adjusting Medication Without Supervision: In a scenario where medical advice is scarce, the ability to adjust your medication based on symptoms and monitor your results is invaluable. For those with hypertension, understanding the relationship between salt intake, stress, exercise, and blood pressure levels can guide adjustments in medication dosage. It's important, however, to have a foundational understanding of how your medications work and the signs of over- or under-medication.

Creating an Emergency Health Plan: Detailing an emergency health plan that includes information on your condition, medications, dosages, and any allergies is vital. This plan should be easily accessible to you and those you trust. Additionally, training family members or companions in basic monitoring and care techniques can provide an extra layer of safety.

Leveraging Community and Technology: Building a support network with others who have similar conditions can provide a wealth of shared knowledge and mutual assistance. Technology, too, plays a crucial role, with medical apps and online resources offering guidance on managing chronic conditions in emergency scenarios.

Armed with knowledge, preparation, and the right tools, individuals with chronic conditions can navigate the complexities of post-disaster healthcare with confidence, ensuring resilience and continuity of care even in the most challenging circumstances.

LONG-TERM PLANNING FOR HEALTHCARE IN UNCERTAIN TIMES

Navigating through the unpredictable terrain of post-disaster life demands foresight, especially when it concerns health. The art of long-term planning for healthcare becomes a vital skill, ensuring that those with chronic conditions can continue their treatment uninterrupted, even as the world changes around them. This preparation is not just about gathering supplies; it's about creating a sustainable healthcare strategy that can adapt to evolving circumstances.

Ensuring the Sustainability of Care

Stockpiling Medications: Building a comprehensive medication reserve is the first step toward healthcare self-reliance. The World Health Organization provides guidelines on the proper storage conditions for pharmaceuticals, emphasizing the need to consider factors like temperature, light, and humidity to maintain medication efficacy over time. A well-thought-out stockpile includes not only a sufficient quantity of medication but also the necessary storage solutions to keep them viable.

Exploring Alternative Treatments: In scenarios where access to conventional medications becomes limited, alternative treatments can play a crucial role. From herbal remedies to physical therapies, these alternatives can offer relief and management for a variety of conditions. Researching and understanding these options before they're needed will ensure that you have a backup plan that's both effective and safe.

Education and Empowerment: Knowledge is a powerful tool in long-term healthcare planning. Educating yourself and your loved ones about your condition, treatment options, and emergency procedures ensures that everyone is prepared to manage health needs effectively. This education can include first-aid training, nutritional advice to manage conditions, and workshops on herbal medicine.

Networking for Support: Building connections with healthcare professionals, support groups, and others managing similar conditions provides a network of advice and assistance. These relationships can be invaluable for exchanging information on managing health conditions under duress, sharing resources, and offering mutual support.

Regular Reviews and Adjustments: A plan is only as good as its relevance to the current situation. Regularly reviewing and updating your healthcare plan ensures that it evolves to meet changing needs, whether due to condition progression, medication availability, or the introduction of new treatment options.

By embracing long-term planning for healthcare, individuals with chronic conditions can secure a level of autonomy and assurance in their treatment. This proactive approach enables not just survival but the maintenance of health and well-being, even in the face of uncertainty.

EMERGENCY PROCEDURES

When the unexpected strikes, the difference between a crisis and a manageable situation often lies in the immediate response. Acute medical emergencies, whether a sudden injury or a life-threatening condition, demand swift, confident action. Preparedness for these moments can save lives, transforming potential tragedy into a story of survival and resilience.

Now, let's explore essential emergency procedures. We've providing clear, step-by-step guidance rooted in the protocols of authoritative bodies like the American Red Cross.

ACUTE CARE GUIDES

CPR for Cardiac Arrest: When someone's heart stops, prompt CPR can be the lifeline they need.

1. Check responsiveness. If the person is unresponsive, shout for help. Call emergency services if alone.
2. Place the person flat on their back and kneel beside their chest.
3. Perform chest compressions: Place the heel of one hand on the center of the chest, placing the other hand on top. Lock your elbows and use your body weight to compress the chest at least 2 inches deep, aiming for 100-120 compressions per minute.
4. If trained, alternate 30 compressions with two rescue breaths. If not, continue compressions only.
5. Keep going until emergency services arrive or the person shows signs of life.

Wound Care: Managing severe bleeding quickly can prevent shock and save a life.

1. Apply direct pressure with a clean cloth or bandage.
2. If the bleeding does not stop, continue applying pressure and raise the injured area above the heart if possible.
3. Do not remove the cloth or bandage if it becomes soaked through; add more layers as needed.
4. Secure the bandage with adhesive tape or hold it in place until medical help arrives.

Choking Relief:

1. Ask the person if they are choking and if you can help.
2. Stand behind them, wrapping your arms around their waist.
3. Make a fist with one hand and place it just above the person's navel, thumb side in.
4. Grasp your fist with the other hand and perform a quick, upward thrust.
5. Repeat until the object is expelled or the person can breathe or cough on their own.

These guidelines follow the trusted protocols established by the American Red Cross. They are designed to offer immediate, life-saving interventions in critical moments. Mastery of these procedures can empower you to act decisively, offering hope and help when every second counts. With knowledge, practice, and the courage to step forward, anyone can become a pivotal force in emergencies, embodying the spirit of preparedness and resilience that defines true survival.

COMPREHENSIVE COVERAGE: NAVIGATING THROUGH LIFE'S UNPREDICTABLE MOMENTS

Life's unpredictability demands preparedness that spans the full spectrum of emergencies, from the joyous cries of new life to the sudden shocks of nature's less welcome surprises. This section delves into essential emergency procedures to empower you with knowledge and techniques to confidently face situations such as emergency childbirth, snake bites, and severe allergic reactions. Each scenario is dissected with precision, offering a clear, actionable guide to ensure safety and well-being when every second counts.

Emergency Childbirth: A Guide to Welcoming Life in Extremis

When a baby decides to arrive ahead of schedule, and professional medical help is not immediately available, knowing the basics of emergency childbirth can turn a potentially panic-stricken situation into a controlled and joyous event. Key steps include:

1. **Preparation:** Ensure the mother is comfortable and in a clean environment. Have clean towels and a blanket ready for the baby.
2. **Delivery:** Encourage the mother to breathe deeply and push with contractions. Support the baby's head and shoulders gently as they emerge, making sure the airway is clear.

3. **Aftercare:** Keep the newborn warm and initiate skin-to-skin contact with the mother. Wait for the umbilical cord to stop pulsing before tying and cutting it with a clean instrument.

Snake Bites: Swift Actions to Mitigate Danger

Encounters with venomous snakes, though rare, require immediate and precise responses to minimize harm:

1. **Immobilization:** Keep the bitten limb as still as possible to slow the spread of venom.
2. **Avoidance:** Do not attempt to suck out the venom or apply ice. These actions can worsen the situation.
3. **Seek Help:** Call for emergency medical assistance immediately, keeping the victim calm and still until help arrives.

Severe Allergic Reactions: Lifesaving Interventions

Severe allergic reactions, or anaphylaxis, can escalate quickly, making rapid recognition and response crucial:

1. **Identification:** Watch for signs of a severe allergic reaction, including difficulty breathing, swelling, and hives.
2. **Epinephrine:** If the person has an epinephrine auto-injector, administer it immediately, following the device's instructions.
3. **Emergency Services:** Call for medical help without delay. Even if symptoms seem to improve, professional assessment is essential.

Equipping yourself with the knowledge to manage these scenarios provides not just the tools for emergency response but also the confidence to act decisively. Preparedness transcends mere survival; it's about preserving the quality of life, sometimes even creating it, and ensuring that when faced with the unexpected, you're never powerless.

ALTERNATIVE THERAPIES

In the quest for health and healing, stepping beyond the conventional medical model opens a world rich with diverse practices known as alternative therapies. These methods, rooted in ancient traditions and validated by modern research, offer a complementary approach to wellness, especially when traditional healthcare options may be limited or unavailable. This exploration into non-conventional medicine not only broadens our toolkit for maintaining health but also deepens our connection with holistic healing practices that have supported human well-being for centuries.

EXPLORING NON-CONVENTIONAL MEDICINE

Alternative therapies encompass a wide range of practices, from acupuncture and massage to energy healing, each with its unique history and methods of addressing physical and emotional ailments.

- **Acupuncture**: With its origins in traditional Chinese medicine, acupuncture involves the insertion of thin needles into specific points on the body. This practice is designed to balance the flow of energy or life force—known as qi (chi)—believed to flow through pathways in your body. Modern research, including studies cited by the National Institutes of Health, has demonstrated acupuncture's efficacy in treating conditions like chronic pain, migraines, and stress.

- **Massage Therapy**: Beyond its well-known capacity for relaxation and stress relief, massage therapy offers significant health benefits, including pain reduction, improved circulation, and enhanced immune function. Historical texts and recent studies alike acknowledge its therapeutic potential, with research highlighted by the American Massage Therapy Association pointing to its role in managing chronic back pain and anxiety.

- **Energy Healing**: Practices such as Reiki and qi gong focus on manipulating the body's energy flow to promote healing. While these therapies might seem esoteric, they are grounded in the belief that physical health is deeply intertwined with the body's energetic state. Publications in peer-reviewed journals have begun to explore the measurable impacts of energy healing on post-operative recovery and chronic pain management, offering a scientific basis for these ancient practices.

These alternative therapies, supported by both historical context and contemporary research, present valuable options for enhancing health, particularly in situations where conventional medical resources are scarce. They remind us that healing is a multifaceted journey, one that benefits from a holistic approach encompassing the wisdom of both ancient traditions and modern science.

PRACTICAL APPLICATION OF ALTERNATIVE THERAPIES

Integrating alternative therapies into daily life offers a pathway to enhanced well-being, especially when traditional medical treatments are out of reach. This practical application is not merely about adopting new practices; it's about weaving these therapies into the fabric of your health strategy, creating a comprehensive approach that values prevention, healing, and resilience. Let's navigate how these therapies can be applied practically, enriching our physical and mental health landscapes.

Applying Alternative Therapies

- **Self-Administered Techniques**: Many alternative therapies can be practiced independently, making them perfect for situations where professional healthcare is unavailable. Simple acupressure techniques can be learned and applied to oneself to relieve pain or stress. Similarly, guided meditation and basic yoga poses can significantly improve mental well-being and are easily integrated into daily routines.

- **Community-Driven Practices**: In many cultures, healing practices are community-centered, offering a model for how we might organize community health initiatives. Group meditation sessions, yoga classes, or even community massage therapy workshops can foster not only individual health but also collective well-being. These activities not only provide direct health benefits but also strengthen community bonds, essential in post-disaster recovery scenarios.

- **Education and Resource Sharing**: Empowering oneself and others through education on alternative therapies enhances the community's overall resilience. Sharing resources, such as books, online courses, or workshops led by knowledgeable practitioners, can help disseminate valuable skills and information. Jones (2021) highlights the role of community-led initiatives in reducing stress and promoting physical health through shared learning experiences.

- **Considerations for Integration**: When incorporating alternative therapies into your health plan, consider their compatibility with existing treatments and conditions. Consulting with healthcare professionals or trusted sources can provide guidance on which practices may be most beneficial for specific health needs.

This chapter emphasized the importance of enhancing medical preparedness as a continuous journey of learning, practicing, and refining skills. It encouraged expanding knowledge, incorporating health strategies into preparedness plans, and committing to ongoing learning and adaptation for personal and communal well-being. The forthcoming chapter promises to explore advanced navigation and communication techniques, aiming to augment survival skills, build resilient communities, and prepare individuals to face future challenges confidently and skillfully.

Medical preparedness is a key aspect of survival. Additional health and
wellness topics like Herbal Medicine and an expanded Prepper's Medical
Handbook are part of the 10 complimentary bonuses you'll receive when
you scan this QR code with your smartphone.

Alternatively, you can visit this website:

https://rebrand.ly/scan-for-your-bonuses

CHAPTER 14
ADVANCED SURVIVAL TIPS AND TRICKS

"Survival is the art of staying alive in the heart of the wilderness; it requires wisdom, resilience, and the will to overcome."

— *Anonymous*

Cap off your journey to ultimate preparedness with "200 Survival Tips and Tricks," a chapter that distills the essence of survival wisdom into actionable advice for overcoming the most challenging environments and situations. From scorching deserts to freezing arctic tundras to dense tropical forests, discover specialized tips that will ensure your survival in extreme conditions. Learn advanced fire-starting techniques that defy wet weather and the absence of matches, ensuring you can create warmth and cook food under any circumstances. Beyond the physical aspects of survival, this chapter also addresses the critical component of mental health maintenance, offering strategies for preserving psychological well-being in isolation or after traumatic events. Emphasizing resilience and the power of community support, we aim to equip you with a comprehensive toolkit for thriving in adversity so you can navigate future uncertainties with confidence.

EXTREME ENVIRONMENT SURVIVAL

Surviving in extreme environments demands more than just courage; it requires knowledge, adaptability, and a keen understanding of nature's rules. Each landscape, whether scorching desert, icy arctic, or dense tropical forest, presents unique challenges and survival strategies. This section breaks down survival into three critical environments, offering actionable tips to navigate and endure these harsh conditions.

DESERTS

The desert, with its vast expanses of sand, relentless sun, and scarce water, tests the limits of survival. Here, every decision could mean the difference between life and death.

Finding Water: Water is life, especially in the desert. Techniques like using solar stills, which involve digging a hole, placing a container at its base, and covering it with plastic sheeting to collect condensation, can be a lifesaver. Similarly, collecting morning dew with a cloth and squeezing it into a container can provide crucial hydration.

Avoiding Heatstroke: The desert sun is unforgiving. Wearing loose, light-colored clothing reflects sunlight, while a wide-brimmed hat shields your face. Resting in the shade during the peak heat (typically between 10 a.m. and 4 p.m.) and staying hydrated are crucial steps in preventing heatstroke. National Geographic Society experts underline the importance of staying out of direct sunlight during these hours to conserve energy and water.

Navigating: Desert navigation relies on both traditional and modern techniques. At night, the stars provide a reliable guide, with constellations and the North Star offering direction. During the day, understanding the landscape's natural indicators, such as the direction of dune formations, can aid in navigation. A compass, if available, is invaluable for maintaining a straight path in featureless terrains.

Surviving in the desert requires a deep respect for its challenges and thorough preparation to meet them head-on. By mastering these survival strategies, one can learn to navigate the desert's beauty and dangers with confidence.

ARCTIC CONDITIONS

The Arctic, a mesmerizing landscape of ice and snow, poses a stark contrast to the desert's heat. The cold does not whisper; it howls and will challenge every bit of your survival knowledge.

Maintaining Body Heat: The first rule of Arctic survival is to stay dry. Moisture wicks away body heat rapidly, a situation you can't afford in freezing temperatures. Dress in layers to trap warm air close to your body, utilizing materials that retain insulation when wet, such as wool. Windproof and waterproof outer layers are essential for protection against the elements. The Arctic Health Research Center advises the importance of these protective layers in preventing heat loss.

Building Snow Shelters: In the vast white wilderness, a snow shelter can mean the difference between life and death. A well-constructed quinzhee or snow cave can shield you from the biting winds and insulate against the cold. Start by piling snow into a mound, and then hollow it out, ensuring the walls are evenly thick to maintain structure and warmth.

Recognizing Signs of Frostbite and Hypothermia: Vigilance for frostbite and hypothermia is vital. Frostbite manifests in numbness and pale, hard skin, primarily on extremities. Hypothermia symptoms include shivering, slurred speech, and confusion. Immediate actions, such as seeking shelter, warming the body slowly, and staying hydrated, is crucial for survival.

TROPICAL FORESTS

The tropical forest, dense and teeming with life, offers a survival challenge cloaked in greenery. The environment is as beautiful as it is demanding, requiring unique strategies to navigate and endure.

Dealing with Heavy Rainfall: Rainfall is a constant in the tropics, making waterproof gear indispensable. Constructing raised shelters and sleeping platforms can keep you dry and protect against ground-dwelling creatures. Channeling rainwater for drinking using tarps or large leaves can also turn the deluge to your advantage.

Navigating Dense Vegetation: The thick underbrush and towering trees of the tropical forest make navigation challenging. Creating clear physical markers as you travel can prevent disorientation. Relying on natural landmarks, such as rivers or mountain ranges visible above the canopy, helps maintain a sense of direction.

Avoiding Tropical Diseases: The lush ecosystem is a breeding ground for diseases. Protective clothing and insect repellent are the first lines of defense against mosquitos carrying malaria or dengue fever. It is a necessity to understand local flora and fauna to avoid plants and animals that could cause harm or illness (Smith et al.).

Surviving in these extreme environments—Arctic cold and tropical forests— demands respect, preparation, and a deep understanding of the unique challenges they present. Mastery of these survival techniques will equip you to face the extremes of our world.

ADVANCED FIRE STARTING

Fire is the beacon of survival, providing warmth, protection, and a means to cook food. Yet, igniting a flame under adverse conditions tests the mettle of even the most seasoned adventurers. This section delves into advanced techniques for fire starting, tailored to overcome the hurdles of wet weather and the absence of conventional tools like matches.

WET WEATHER

Damp conditions can turn fire-starting into a daunting task. Moisture in the air and on the ground snuffs out weak flames and makes tinder reluctant to catch. However, with the right knowledge, even the wettest environments can yield the warmth of a fire.

- **Using Resinous Wood**: Trees like pine contain resin, which can burn even when wet. Look for dead branches or stumps, where resin accumulates, and use shavings or chunks of this wood as part of your tinder bundle.
- **Creating a Fire Platform**: Elevating your fire from the wet ground is crucial. Construct a platform using larger sticks and logs, ensuring it's sturdy enough to hold your fire. This barrier prevents ground moisture from reaching your tinder and allows air to flow underneath, enhancing combustion.

WITHOUT MATCHES

The absence of matches demands a return to the basics, utilizing friction, sunlight, or chemistry to spark a flame.

- **Bow Drill Method**: This age-old technique requires a bow, spindle, fireboard, and bearing block. The bow is used to rotate the spindle rapidly against the fireboard, generating heat through friction to create an ember. Proper form and patience are the key to success with this method.
- **Fire Plough**: Similar to the bow drill, the fire plough relies on friction but involves rubbing a hard stick along a groove in a softer wood base. The motion and pressure generate heat, culminating in the formation of embers along the groove's path.
- **Using a Lens to Focus Sunlight**: On sunny days, a lens can concentrate sunlight onto a specific point in your tinder, raising its temperature to ignition. This method works with eyeglasses, magnifying glasses, or any transparent material capable of focusing light.

Mastering these advanced fire-starting techniques ensures that you can create warmth and light in any situation, turning the challenge of ignition into an achievable task. Whether faced with a deluge or devoid of modern conveniences, the ability to start a fire remains a pivotal survival skill.

MENTAL HEALTH MAINTENANCE

In the wilderness of survival, the landscape of the mind requires as much navigation as the physical world. The psychological rigors of survival situations bring unique challenges, where the maintenance of mental well-being becomes as critical as securing water, food, and shelter. This section addresses the mental fortitude needed to thrive, not just survive, introducing practices and strategies designed to fortify the mind against the pressures of uncertainty and isolation.

STRATEGIES FOR WELL-BEING

In the silence and solitude of survival, the mind can become one's greatest ally or most formidable foe. Activities that foster mental health are essential tools in the survival kit.

- **Mindfulness and Meditation**: These practices anchor the mind in the present, reducing stress and enhancing resilience. Simple techniques, such as focused breathing or mindful observation of nature, can offer solace and clarity. Greenberg (2021) highlights the profound impact of mindfulness on improving mental health in isolation, providing a pathway to inner peace amidst external chaos.

- **Journaling**: Writing not only documents the journey but also allows you to process emotions and experiences. It can serve as a reflective practice, allowing you to navigate thoughts and feelings, set goals, and celebrate successes, no matter how small.

- **Establishing a Routine**: The unpredictability of survival scenarios can erode one's sense of control. Establishing a daily routine, as advised by Greenberg, brings structure and normalcy, significantly improving mental health. Whether it's dedicated times for meals, exercise, or rest, a routine can anchor the day and provide much-needed stability.

RESILIENCE AND COMMUNITY SUPPORT

The human spirit's resilience can be amplified through positive thinking and the power of community. Even in isolation, a sense of connection to others—whether through memories, written letters, or anticipation of reunion—can sustain the heart.

- **Cultivating Positive Thinking**: Acknowledging the difficulty of circumstances while focusing on what can be controlled can foster a mindset of resilience. Visualization of positive outcomes and gratitude practices can shift perspectives, lighting the path through dark times.

- **Leveraging Community Support**: In situations where others are present, building a supportive community can become a lifeline. Sharing stories, skills, and tasks can strengthen bonds, distribute the psychological load, and help you foster a shared resilience. The collective spirit of a group can conquer challenges far beyond the reach of an individual.

Maintaining mental health in survival situations requires conscious effort and the implementation of strategies that nurture the mind. By embracing mindfulness, journaling, routine, positive thinking, and the strength of community, you can equip yourself with the psychological resilience necessary to face any adversity.

Thank you for purchasing my book, ""The Complete Prepper's Survival Bible." I want you to have the prepping knowledge you need at your fingertips to safeguard your present and future. When you scan the QR code above with your smartphone, you'll unlock 10 valuable bonuses that will help you in your journey. Alternatively, you can visit this website: https://rebrand.ly/scan-for-your-bonuses

CONCLUSION

"Survival is not the art of living; it's the art of existing when
everything else has failed."

This wisdom, attributed to an anonymous sage of survival, captures the spirit
we've navigated through these pages. From laying down the basics of
preparedness to exploring the depths of self-reliance, this guide has aimed to
equip you with not just the tools but the mindset to thrive in adversity.

Preparedness begins with understanding—the knowledge that the
unexpected is part of our world. We've delved into how crucial it is to have a
plan, the right supplies, and the skills to use them. It's not about having a
stockpile of goods but about knowing what truly matters: water, food,
shelter, and the ability to ensure your safety and health.

As we progressed, we explored beyond the basics into sustainable living, from
growing your own food to harnessing natural resources for energy.

Summary of Key Ideas

Prepping Foundations: At the core of preparedness lies a simple truth:
readiness begins with a solid foundation. We've unpacked the essentials of
prepping, emphasizing not just what to have but how to think. The right
mindset, coupled with careful planning and a grasp of the basics, sets the stage
for a resilient life. Whether it's securing your water source, understanding the
nuances of food storage, or mastering the art of staying warm and safe, these
chapters are your first steps toward self-reliance.

Advanced Skills and Knowledge: Beyond the basics, the journey into preparedness deepens. We've ventured into the realms of sustainable living, medical preparedness, and the art of growing and preserving your own food. Each skill, each piece of knowledge, adds layers to your ability to thrive in any circumstance. This book serves as a reminder that learning is a never-ending process. By embracing advanced practices, you're not just preparing for emergencies; you're crafting a lifestyle that's both resilient and rewarding.

Ethical and Community Considerations: Preparedness is more than a personal journey; it's a commitment to ethical living and community building. We've explored how responsible resource management and respect for our environment are crucial. By engaging with and supporting those around us, we can forge networks of resilience that extend beyond our individual capabilities. This book advocates for a preparedness approach that strengthens not just our homes but our communities, emphasizing the power of collective effort and mutual support.

Personal Growth and Resilience: The path to preparedness is also a path to personal growth. Mental toughness and emotional resilience are the unsung heroes of survival. We've delved into the psychological aspects of facing challenges, highlighting how these inner strengths are essential. Building resilience isn't just about surviving; it's about thriving in the face of adversity, equipped with the confidence to tackle any challenge.

Alternative Approaches: Embracing diversity in our strategies, we've considered alternative therapies and unconventional solutions. This exploration underscores the importance of keeping an open mind and being willing to adopt a broad perspective on problem-solving and well-being. From herbal remedies to innovative off-grid solutions, the array of options available enriches our preparedness toolkit, offering multiple paths to a secure and healthy life.

Through these chapters, the message is clear: Preparedness is a multifaceted endeavor. It's about building a foundation, expanding our skills, embracing our community, growing personally, and exploring all avenues to resilience. This journey equips us to not only face the future with confidence but to shape it with our actions today.

Now is the time to turn insight into action. Begin with small, achievable steps: assemble an emergency kit, start a vegetable garden, or learn a new survival skill. These initial actions are the building blocks of a broader, more secure future. Remember, preparedness is not a destination but a journey of lifelong learning. Commit to staying informed, regularly honing your skills, and revising your plans as the world around you evolves.

Beyond personal readiness, extend your hand to the community. Connect with others who share your commitment to resilience. Share your knowledge, learn from others, and build a network of mutual support. Together, you can create a fabric of preparedness that not only weathers storms but strengthens the bonds between you. Community engagement multiplies your resources, broadens your safety net, and reinforces that in unity there is strength. As you move forward, remember: every small step you take is a leap toward a future where you stand ready, resilient, and connected.

Preparedness transcends mere survival; it's about flourishing in the face of challenges and being empowered to meet the future with confidence. This journey will equip you with more than skills and supplies; it will foster a mindset of resilience and adaptability. As you close this book, carry forward the belief in your ability to thrive amidst adversity. Let preparedness be your beacon, guiding you not just through storms but toward a brighter, more secure future. Remember, the act of preparing is itself a profound expression of hope and a testament to the human spirit's capacity to prevail.

Made in the USA
Coppell, TX
02 October 2024

37981248R10116